THE GREAT INDIAN NATURE TRAIL

WITH
CHUNMUN

ROHAN CHAKRAVARTY
WITH ESSAYS BY BIJAL VACHHARAJANI

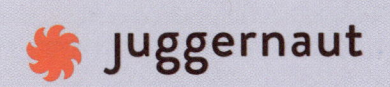

JUGGERNAUT BOOKS
C-I-128, First Floor, Sangam Vihar, Near Holi Chowk,
New Delhi 110080, India

First published by Shashthi Media and Juggernaut Books 2024

Copyright © WWF-India 2024

10 9 8 7 6 5 4 3 2 1

P-ISBN: 9789353458591
E-ISBN: 9789353452155

All rights reserved. No part of this publication may be reproduced, transmitted, or stored in a retrieval system in any form or by any means without the written permission of the publisher.

Disclaimer: The maps in the book are not to scale and do not depict the political boundaries accurately.

The comics were created under the 'One Planet Academy' project that is supported by Capgemini India. One Planet Academy (https://academy.wwfindia.org/) is a digital resource centre for environment education for students and educators.

Capgemini

Printed at Thomson Press India Ltd

MIX
Paper | Supporting responsible forestry
FSC® C010615

This book is produced from independently certified FSC® paper to ensure responsible forest management.

Explorers, welcome back!

Or if this is the first time you are getting aboard The Great Indian Nature Trail, then, welcome.

The Great Indian Nature Trail took place with Uncle Bikky, his dog Duggu and his niece, Chunmun who, in part one, was a budding wildlife photographer. Since then, Chunmun has grown up and is now the youngest person in India to win a grant from the *Kingdom Animalia* magazine to work on a wildlife photo essay.

She gets to visit the Western Ghats, the North-east of India, the tide pools and rivers of this country and the forests. She gets to meet some amazing people on her adventures – scientists, naturalists, filmmakers and researchers. All of them introduce her to a different facet of wildlife.

And you get to join her!

Congrats, you've won a virtual ticket to join Chunmun on her wild expedition across India.

So what are you waiting for? Pack your bags, don't forget to add a pair of binoculars, you are going to need that. And maybe some... wait, one of the most carbon-friendly kind of travel is the one that you are going to do – by enjoying a staycation (a vacation where you stay at home) and read this book.

Happy adventuring!

Rohan and Bijal

BUCKLE UP!

Where all is Chunmun travelling? Here's a sneak peek.

WHO ALL DOES CHUNMUN MEET?

Here are some of the species you will meet in the book. Once you do meet them, don't forget to tick them off here. Bonus points if you've met them in real life.

Chestnut-breasted Partridge ....

Bugun Liocichla ....

Collared Owlet ....

Sclater's Monal ....

Hoolock Gibbon ....

Indian Tent Turtle ....

Gangetic Dolphin ....

Hangul ....

Striped Hyena ....

Rusty-spotted Cat ....

Indian Leopard ....

Fishing Cat ....

Greenish Warbler ....

Indian Skimmer ....

Gharial ....

Indian Humpback Dolphin ....

Fungoid Frog ....

Malabar Gliding Frog ....

Indian Purple Frog ....

Owlfly ....

Antlion ....

Ant-mimic Spider ....

Sea Anemone ....

Sand Bubbler Crab ....

Hermit Crab ....

Seastar ....

Amur Falcon ....

Crested Kingfisher ....

Brown Dipper ....

A very special bird you're rather unlikely to ever meet! ....

5

1
FROM THE NORTH-EAST, WITH LOVE

WOOF! WRRUFFF!!

GOOD MORNING POSTMASTER DUGGU! WHAT HAVE YOU GOT THERE?

A PARCEL FROM CHUNMUN, ALL THE WAY FROM ARUNACHAL!

TO, MR. BIKASH GANGULY

FROM CHUNMUN GANGULY

HEY, OLD PAL! HOW IS THE ARTHRITIS TREATING YOU?

REMEMBER OUR TRIP TO UTTARAKHAND WHEN WE TREKKED HOPING TO REDISCOVER THE HIMALAYAN QUAIL, BUT FLUSHED OUT A HILL PARTRIDGE INSTEAD?

KURRRRR!

WELL, LAST WEEK I SPOTTED A VERY SIMILAR BIRD — THE CHESTNUT-BREASTED PARTRIDGE, A RARE RESIDENT OF THE NORTH-EAST HIMALAYAN FORESTS...

AND GUESS WHAT, IT CALLS LIKE THE HILL PARTRIDGE TOO!

KRROOOAAA!

UNCLE BIKKY, I DIDN'T TELL YOU ABOUT THIS EARLIER. I WANTED TO SURPRISE YOU —

I'VE WON THE KINGDOM ANIMALIA EXPLORER'S GRANT FOR PREPARING A PHOTO ESSAY ON INDIAN WILDLIFE! I'M THE YOUNGEST PERSON IN INDIA TO WIN THIS GRANT!

FOR MY PROJECT, I WILL BE TRAVELLING AROUND INDIA, PHOTOGRAPHING, WRITING AND LEARNING ABOUT INDIA'S BIODIVERSITY!

THIS REALLY WOULDN'T BE POSSIBLE WITHOUT ALL THE TREKS, HIKES AND TRAILS YOU'VE TAKEN ME ON...

SO HERE'S A LITTLE GIFT FOR YOU FROM THE FIRST WING OF MY EXPEDITION — SOME FIERY 'BHOOT JOLOKIA' RED CHILLIES TO MAKE YOU WAIL LIKE THE HILL PARTRIDGE!

AND SOME WILD TURMERIC TO CURE YOUR KNEES AND BRING YOU BACK FROM 'EXTINCTION' LIKE YOUR FAVOURITE BIRD, THE HIMALAYAN QUAIL!

Isn't Chunmun's work amazing? She gets to travel across India, photographing and learning about the country's biodiversity.

WHAT DOES A WILDLIFE PHOTOGRAPHER DO?

Of course, they get to travel to tonnes and tonnes of places – from forests to rivers to deserts and even to the ice caps. But it's not just all travelling. Photography is a lot of work and responsibility. They look for stories through their camera, tell stories with their images, and we all know...
a picture is worth a thousand words.

With their images, photographers have told us about the way animals live in the wild – their social behaviour and feeding habits. They have taken us to different parts of the world to see habitats such as rainforests and cloud forests, grasslands and mangroves, wetlands and deserts. Photographers have acquainted us with fungi and microbes, carnivorous plants and algae; also, made us care about issues such as poaching, climate change and habitat degradation. They have aided in the constant discovery that the Earth is a unique home to live in, and we are lucky to share it with such amazing flora and fauna.

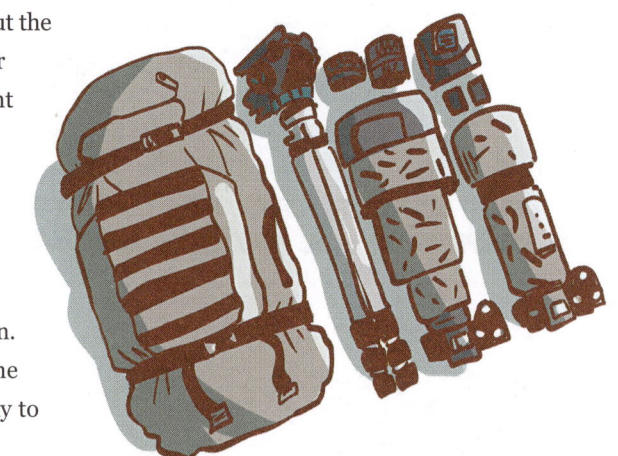

BEYOND THE SHOT

The first thing that wildlife photographers have to keep in mind is that they do need to get the best shot, of course. Excuse me, a film shot! What did you think? It's the only shot we are talking about in this book. Unless we are condemning poaching.

Yes, back to shooting, of the right kind. Photographers need to get the best shot without impacting the fragile habitat they are in. In fact, that's so important that there are even unwritten rules for wildlife photography. Such as, don't use lights at night for that perfect shot of an owl or a bat. This is easy to figure, right?

Then, no nest photography. We all love fledglings and their open beaks, forever hungry for food. But these nests are built by birds with a lot of care to ensure predators can't get to them. Sometimes in their zeal to get a perfect shot, photographers may disturb the area around the nest, for example, by clearing crucial foliage, and whammm, they serve up a poached egg for predators.

Put yourself in the shoes of...
a wildlife photographer!
- What would your day look like?
- How would you help wildlife with your photos?
- Most importantly, how would you ensure you won't harm wildlife with your camera?

SOME AWESOME WOMEN WILDLIFE PHOTOGRAPHERS AND FILMMAKERS

Munmun Dhalaria: That's right, the inspiration behind Chunmun is a real-life filmmaker. Her films include *The Jujurana's Kingdom*, *On the Brink* and *COVID Response*.

Arati Kumar-Rao: Ever met an explorer? That's who this environmental photographer is, she's a *National Geographic* explorer and she writes, draws and takes photos!

Ami Vitale: The North American photojournalist has taken some amazing photos of elephants and pandas. Don't miss them.

Ashima Narain: Bears, flamingos and leopards are just some of the subjects of Ashima's camera.

DRAW

Can you imagine Uncle Bikky wailing like the Hill Partridge? Now go listen to the bird's call, and then, learn how to draw the bird, wail and all.

FUN FACTS

1. The Chestnut-breasted Partridge's Latin name, *Arborophila mandellii*, is said to commemorate the Italian naturalist Louis Mandelli, who lived and worked in Darjeeling.
2. This partridge, like many other partridges and pheasants, has spurs at the back of its feet, which males use in combat with rivals!
3. The 'Bhoot Jolokia' (which translates to 'Ghost Pepper'), grows in the hills of North-east India, and is one of the hottest chillies in the world!
4. Wild turmeric is widely believed to have antiseptic and therapeutic properties, and is used to treat a variety of ailments in Ayurveda.
Q. Which part of the turmeric plant does 'haldi powder' come from?

HUNGRY FOR MORE?

Did you see that Chunmun sent some bhoot jolokia chillies for Uncle Bikky? It has the distinction of being one of the spiciest naturally-occurring chillies in the whole wide world, and it's grown right here in the North-east of India.

Wait, how is spiciness measured? Is it by downing different chillies and seeing which one makes you weep the most?

No way!
There's something called the Scoville Heat Units (SHU), which is a way of measuring spiciness by the concentration of capsaicin (that's the chemical that has you reaching for water when you eat something spicy). The bhoot jolokia or ghost pepper has an average rating of 1,041,427 SHU.

GROW A CHILLI AT HOME

It's really easy! Raid your fridge. Take a chilli, carefully slice it and remove the seeds from the inside. Be careful not to touch your eyes or any part of your face! Don't forget to wash your hands properly.

Dry them for a couple of days. Sow them in soil, a couple of centimetres deep. Now water it every two days. The seeds should germinate in a week or so. In a couple of months, get ready to see beautiful white flowers. Once they wither, it's the chilli's time to bloom. Here's a place to paste the photo of your cool but hot chilli plant.

Riddle Dee Dee like the rhyme
They don't laugh back when you tell plants a joke, but if you choose a windy day, they will shake (with laughter). If you're not the funny sort, sing a song or just chat to the plant.

In a study,[1] researchers found that plants can sense sound vibrations – such as running water – and then their roots inch towards the water source! In fact, plants move away from noises they don't like. So, if the leaves start looking away, blame it on your jokes.

Don't forget to talk to your chilli plant and take care of it. Jokes like these are much appreciated
Knock knock,
Who's there?
Chilli.
Chilli who?
It's chilli out here, open the door quickly.

2. BUGUN IN THE BUSH

CHUNMUN'S FIRST ASSIGNMENT FOR HER PHOTO ESSAY TAKES HER TO EAGLENEST WILDLIFE SANCTUARY IN ARUNACHAL PRADESH. HERE SHE ACCOMPANIES ORNITHOLOGIST DR UMESH ON A FIELD VISIT.

HOW MANY TIMES HAVE YOU SEEN THE BUGUN LIOCICHLA, UMESH?

WOW, CHUNMUN, I'M VERY SURPRISED THAT YOU EVEN KNOW ABOUT THE BUGUN LIOCICHLA!

A BIRD SO RARE THAT IT'S ONLY FOUND IN THIS TINY PATCH OF RAINFOREST IN THE WHOLE WORLD! THERE MAY BE FEWER THAN A HUNDRED!

TO STUDY HOW HILL BIRDS MOVE FROM ONE ALTITUDE TO ANOTHER, I CATCH THEM IN THESE MISTNETS, AND TAG THEM WITH ALUMINIUM RINGS...

BUT THE BUGUN LIOCICHLA KEEPS ELUDING ME!

HUH! WHAT'S THAT COMMOTION?

OH MY GOSH! I'VE NEVER SEEN SO MANY SPECIES OF BIRDS TOGETHER! WHAT'S GOING ON?
THIS IS CALLED A BIRD WAVE! BIRDS SUCH AS DRONGOS, BABBLERS, TITS, WARBLERS, NUTHATCHES, WOODPECKERS AND PARROTBILLS MOVE AND FEED TOGETHER. THIS HELPS THEM FIND FOOD WITH GREATER SUCCESS, AND ALSO KEEP BETTER VIGIL ON THEIR PREDATORS...
POO-PUPUP-OOOP!
AH! SPEAKING OF PREDATORS...
POO-PUPUP-OOOP!
A COLLARED OWLET! THIS TINY OWL IS A DREADED PREDATOR OF BIRDS HERE IN THE EASTERN HIMALAYAS. EACH TIME IT CALLS, SECRETIVE BIRDS SCUTTLE OUT OF THE BUSHES IN FRIGHT!
THERE THEY ARE! A PAIR OF BUGUN LIOCICHLAS SCURRYING AWAY!
CLICK
CLICK
CAN YOU IMAGINE, NOBODY KNEW ABOUT THIS SPECIES UNTIL 2006, WHEN DR RAMANA ATHREYA DISCOVERED THE BIRD, AND NAMED IT AFTER THE BUGUN TRIBE OF ARUNACHAL?

> I COULDN'T HAVE ASKED FOR A BETTER START TO MY PROJECT! THANK YOU, UMESH!
>
> WELL, IT'S THE BUGUN PEOPLE WHO YOU SHOULD THANK FOR PROTECTING THE HOME OF THE BUGUN LIOCICHLA...
>
> POO-PUPUP-OOOP!
>
> AND NOT TO FORGET, OUR FRIEND, THE COLLARED OWLET!

FUN FACTS

1. The Bugun Liocichla is not the only species endemic to Eaglenest. A species of frog, the Bompu Litter Frog, is also found only in the bamboo forests of this wildlife sanctuary!
2. The Bugun people are one of India's earliest recognized tribes. They are traditionally animists, and worship the forces of nature.
3. Eaglenest Wildlife Sanctuary derives its name from the Red Eagle Division of the Indian Army. In the 1960s, the sanctuary was a battleground for the Sino-Indian war!
4. Eaglenest is also home to many other rare birds such as the Blyth's Tragopan and mammals such as the Red Panda.
Q. A species of monkey was also discovered in Eaglenest. Can you name this species?

TOGETHER WE CAN

The story of the Bugun Liocichla is amazing. Indigenous[1] communities from across the world are the protectors of these precious fauna, flora and their land. In fact, research shows that it's indigenous people and local communities that have the best conservation solutions and outcomes.

MEET NEWBIES ON THE BLOCK

The Earth is so vast that we are constantly stumbling upon and discovering new species on the planet. Dr Ramana Athreya is the one who discovered the **Bugun Liocichla**.

The *Eriovixia gryffindori* or the Sorting Hat spider was discovered in the Kans region of the Western Ghats of Karnataka. Discoverers: Rajashree Khalap, Javed Ahmed and Sumukha Javagal.

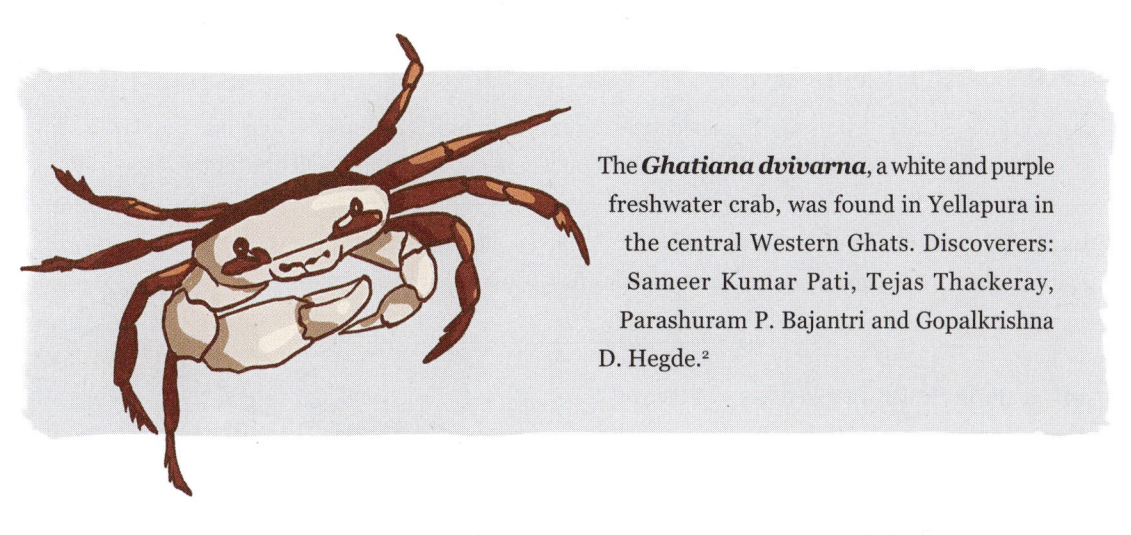

The **Ghatiana dvivarna**, a white and purple freshwater crab, was found in Yellapura in the central Western Ghats. Discoverers: Sameer Kumar Pati, Tejas Thackeray, Parashuram P. Bajantri and Gopalkrishna D. Hegde.[2]

Zeeshan A. Mirza and team discovered the **Hemidactylus vijayraghavani**, a ground-dwelling gecko in Karnataka. He also discovered the **Trimeresurus salazar**, a green pit viper in Pakke Tiger Reserve[3] in Arunachal Pradesh. Also, Zeeshan has discovered some 35 species to date![4]

NAME FAME

Herpetologist Ashok Captain has two snakes named after him! Captain's Wood Snake and Ashok's Bronzeback Tree Snake.

BIRD WATCHERS ANONYMOUS

Everyone knows that one person who can identify birds by just seeing a silhouette or shadow and will tell you it's so and so or that and that. Here are some tips that will make you a top notch ornithologist.

Observe the bird's beak – it's the beak that tells you about its diet. Like the kingfisher's beak is perfect for catching and spearing fish, while a sparrow's conical beak makes it easy to eat grain.

Look at their claws – webbed feet will tell you it's a wader bird, for instance.

Check out their colouring, or whether it has a crest or whiskers; all of them add up to telling you more about the bird.

Listen closely to their calls, and soon you will be able to distinguish one bird from the other. The website *Xeno Canto*[5] has bird sounds to listen to for free.

DID YOU KNOW?

Twitchers are birdwatchers who have a long list of rare birds to see and want to tick them off their list.

DRAW

Learn how to draw the Collared Owlet in four easy steps.

ICYMI (IN CASE YOU MISSED IT)

Researchers Umesh Srinivasan, Nandini Velho and forest officer Millo Tasser, are some of the people helping protect the Bugun Liocichla (Source: *The Wire*).[6] Not to forget the Buguns, who came together to protect the bird's home.

3
MYSTICAL MISHMI

CHUNMUN TRAVELS TO INDIA'S EASTERNMOST CORNER, THE MISHMI HILLS IN THE DIBANG VALLEY OF ARUNACHAL PRADESH. HERE SHE MEETS MR IPRA, A VETERAN CONSERVATIONIST FROM THE NATIVE MISHMI TRIBE.

CLICK CLICK CLICK

IPRA JI, THIS IS INCREDIBLE! I CAN'T BELIEVE WE STARTED THE MORNING WITH A SIGHTING OF THE ELUSIVE SCLATER'S MONAL!

NOT MANY ARE AS LUCKY, CHUNMUN! I'VE HAD SO MANY BIRDER FRIENDS GETTING ALMOST FROSTBITTEN WAITING FOR THIS BIRD HERE!

MAYODIA PASS 2655m

HOW DO YOU MISHMIS KEEP YOURSELF FROM FREEZING IN THIS COLD?

OH, WE DRINK LOTS OF RICE BEER, AND WE TELL EACH OTHER STORIES!

I CAN'T RECOMMEND RICE BEER TO AN EIGHTEEN-YEAR-OLD, BUT I CAN CERTAINLY TELL YOU STORIES! DIBANG VALLEY IS THE LAND OF FOLK TALES...

AND WE MISHMI PEOPLE HAVE A STORY ABOUT EVERY ANIMAL THAT LIVES HERE!
THE TIGER, FOR EXAMPLE, IS BELIEVED TO BE OUR ELDER BROTHER. HUNTING A TIGER IS CONSIDERED THE MOST SERIOUS CRIME.
THERE'S ONE ABOUT THE RARE AND NOCTURNAL MISHMI FLYING SQUIRREL. ONCE A SQUIRREL WANTED TO CROSS A RIVER AND BEGGED A MAN TO LEND HIM HIS PAIR OF WINGS. SINCE THEN, SQUIRRELS CAN GLIDE BUT MAN CANNOT!
THE MITHUN, A SEMI-DOMESTIC OX THAT YOU'LL SEE HERE FREQUENTLY, IS A VERY SPECIAL ANIMAL TO US, AND A SYMBOL OF OUR CULTURAL IDENTITY.
WE BELIEVE THAT IT WAS THE FIRST ANIMAL TO DESCEND TO EARTH, AND COULD BLOCK THE SUN WITH ITS ENORMOUS BULK!
OOO-HEOOUUWWW
OOO-HEOOUUWWW
OOO-HEEEOUUWWW!!
AH, HOW COULD I FORGET THE HOOLOCK GIBBON, INDIA'S ONLY APE? THERE'S A REALLY FUNNY MISHMI MYTH ABOUT THIS ARBOREAL PRIMATE...
MISHMIS BELIEVE THAT THE HOOLOCK GIBBON GOT ITS CHARACTERISTIC HOWL AFTER TASTING APHUNG – OUR LOCAL RICE BEER!
HA HA HA HA!!
BUT OUR CLOSE RELATIONSHIP WITH WILDLIFE IS UNDER GROWING THREAT.
THERE ARE OVER A HUNDRED DAMS BEING PLANNED IN OUR FRAGILE VALLEY, WHICH WILL SUBMERGE HUGE TRACTS OF PRISTINE FOREST AND THREATEN TRIBAL LIVELIHOODS.

> CHUNMUN, WE MOUNTAIN FOLK BELIEVE IN BARTER. SO, IN EXCHANGE FOR ALL THE STORIES I'VE TOLD YOU TODAY, I'LL ASK FOR A SMALL FAVOUR... WILL YOU TAKE THE VOICE OF OUR LAND TO THE REST OF INDIA THROUGH YOUR WILDLIFE PHOTO ESSAY?

> IPRA JI, I PROMISE YOU – I'M GOING TO MAKE MY PHOTO ESSAY HOWL AS LOUD AS THE GIBBONS OF MISHMI HILLS!

> OOO-HEOOUUWWW OOO-HEOOUUWWW OOO-HEEEOUUWWW!!

FUN FACTS

1. There are twenty species of gibbons in the world, and they're all found in South-east Asia.
2. A bird named after the Mishmi tribe – the Mishmi Wren-Babbler – is also found only in Dibang Valley.
3. The Mishmi people are renowned for building bridges out of bamboo.
4. Around spring, the Mishmi people celebrate a harvest festival called Reh, in which Mishmi folk tales are narrated and performed.
Q. An endangered goat-antelope, whose name rhymes with 'napkin', is also found in the Mishmi Hills. Can you name this animal?

MISHMI AND THE MISHMIS

The Mishmis are an indigenous group who live in Tibet and Arunachal Pradesh. The state of Arunachal is already teeming with biodiversity, and in the Mishmi Hills, you can hope to meet the Sclater's Monal, the Blyth's and Temminck's Tragopans, the Mishmi Wren-Babbler, the Bengal Florican and of course the Mithun and Hoolock Gibbons.

WORD!

The Mishmis have their own spoken language, but it has no written script.[1]

FACT

- Indigenous communities are one of the biggest guardians of biodiversity.
- They are superheroes protecting wildlife and land.

Sorry peacock, when it comes to looking good, you've got some serious competition. Yes, both the Sclater's Monal and Himalayan Monal will give you a run for your money. The Himalayan Monal is a fantastic digger. And the males are known for their fabulous courtship dance.[2] You can find them boogieing in states like Uttarakhand and Sikkim.

Alas, both the Indian Peafowl and the Himalayan Monal are poached. The peacock for its tail feathers and the monal for its crest.

READ

WWF-India's book *Birds in your Backyard and Beyond*, written by Arthy Muthanna Singh, Mamta Nainy and Kaustubh Srikanth with art by Aniruddha Mukherjee.

WRITE YOUR THOUGHTS ABOUT THE BOOK HERE

Name of the book:
Author name:
Year of publication:
Publishing house:
Synopsis (Write in one paragraph what this book is about):

What did you like about it?

What did you not like about it?

Conclusion:

DRAW

Figure out the differences between the two monals by learning to draw them.

Sclater's Monal   Himalayan Monal

WRITE TIME!

Mishmi legend says that the Mithun – the semi-domestic ox – was the first animal to descend to Earth. With its enormous body, it could even block the Sun. Maybe that explains eclipses in lore!

Write an origin story of the Mithun – how did she first come to Earth, and how did she hide the Sun?

DAM IT!

Did you hear that dams are threatening the fragile ecosystem of the Mishmis' home? Here is an image of the Mishmi Hills landscape. Observe how the various forms of life, including humans, are interacting with nature. If a dam were built on the river, how do you think it would affect this picture? Imagine and draw the same place below with a dam:

READ

The Dam by David Almond and Levi Pinfold is inspired by a real-life story. A father and daughter return to a landscape silenced by a dam, and fill it with music.

Rinchin and Sagar Kolwankar's award-winning *I Will Save My Land* tells the story of a girl who fights back when she hears her land may become a coal mine someday.

ICYMI

Ipra Mekola is a member of the Idu-Mishmi tribe. A historian, he quit his job to protect nature as a wildlife conservationist.[3] For his work, he has been honoured with the RBS Green Warrior Award.

4

DOLPHIN DATE

TO PHOTOGRAPH THE ENDANGERED AND ENDEMIC GANGETIC DOLPHIN, CHUNMUN TRAVELS TO VIKRAMSHILA IN BIHAR, INDIA'S ONLY DOLPHIN SANCTUARY, WITH HER IDOL AND MENTOR, MS ARATI, A PHOTOJOURNALIST.

OVER HERE, CHUNMUN! AN INDIAN TENT TURTLE BASKING IN THE OPEN!

PLOP!

!

CLICK

PLOP!

THAT WAS GOOD TIMING, CHUNMUN! BUT OUR TARGET FOR THE DAY IS GOING TO NEED MUCH FASTER REFLEXES TO SHOOT.

IF WE'RE LUCKY ENOUGH, THE GANGETIC DOLPHIN WILL EMERGE ONLY FOR A FRACTION OF A SECOND. OFTEN ALL ONE SEES IS JUST ITS DORSAL FIN!

AND I GUESS THE FACT THAT THERE ARE SO FEW OF THEM ALIVE, MAKES OUR TASK ALL THE MORE DIFFICULT!
CORRECT. ITS NUMBERS HAVE CRASHED DRAMATICALLY BECAUSE OF HUNTING, HABITAT LOSS, DAMS, ENTANGLEMENT IN FISHING GEAR AND POLLUTION OF THE GANGA, WHICH GOES ON UNCHECKED...
ALTHOUGH THE GANGETIC DOLPHIN IS BLIND, ITS SENSES ARE RAZOR-SHARP, AND IT IS PERFECTLY ADAPTED TO HUNT IN THESE SILTY WATERS USING ECHOLOCATION!
ARATI, I FIND IT SO IRONIC THAT DESPITE ALL THE DAMAGE WE'VE INFLICTED ON THE DOLPHIN, IT STILL BEARS ITS CHARACTERISTIC SMILE!
YOU'RE RIGHT! AS IF HAVING ATTAINED NIRVANA!
HEY!
I THOUGHT I JUST SAW A SNOUT APPEAR AND DISAPPEAR IN A FLASH! GET READY...
AND REMEMBER TO SHOOT IN A WIDE ARC ON 'HIGH FRAMES', WITH A GOOD DEPTH OF FIELD... THE DOLPHIN COULD APPEAR ANYWHERE...
ROGER!
CLICK
CLICK

> GREAT TIMING, ONCE AGAIN, GIRL!

> I CAN SEE THAT THE DOLPHIN'S SMILE IS VERY INFECTIOUS!

> HAHAHAHA!

The Gangetic Dolphin whom you just met is also known as the South Asian River Dolphin. A freshwater mammal, the dolphin is in hot water.

Correct, climate change is creating hot water. Literally.

But also, remember the bit that Ms Arati, the photojournalist said? That while the dolphin cannot see, it uses echolocation to navigate its river home.

Echolocation is very cool; animals use it to gauge distance and locate where their prey is. They do it by emitting sound that is reflected by objects and prey, and that's how they can experience and navigate the waters they live in.

FUN FACTS

1. The Gangetic Dolphin is the National Aquatic Animal of India!
2. A subspecies of the Gangetic Dolphin is found in the Indus River, and is called the Indus Dolphin.
3. Vikramshila is also home to the Smooth-coated Otter, the Gharial and a variety of freshwater turtles.
4. The Gangetic Dolphin uses echolocation to navigate and hunt in murky waters, by emitting clicks that get reflected by surrounding objects. Its echolocation patterns are very similar to those of the Amazon River Dolphin!
Q. The east coast of India is home to another species of dolphin that lives both in fresh and brackish water. Can you name this species?

WHO ELSE USES ECHOLOCATION?

Hint
1. They're winged mammals that are seen during the night.
2. Some people are scared of them, but that's just silly because these animals are pest controllers and we do need them hanging around.
3. They love to hang upside down.

Answer: bats

HELP CHUNMUN WRITE A RAP ON BATS

There are these mean, ill-informed people who are constantly petrified that bats spread viruses in humans. Nope. They don't. And in 2020, over 60 researchers came together to say to policy makers, that there is no evidence that humans contract viruses through bat excreta.[1] So, write a poem praising bats. Here are some facts you can add to your poem:
- Their scientific name, Chiroptera, means 'hand wing' in Greek!
- Bats eat pests, and so are excellent pest controllers.
- They are good listeners – in fact, some species can even hear a beetle walking![2]

MY POEM

....

....

....

BUT IF ECHOLOCATION IS SO AMAZING, THEN WHAT'S AILING THE DOLPHIN?

Well, have you taken a look at our rivers? Go on, look closer. You will find lots of boats bobbing about. Big ones, small ones – it's so full that there's little place for these freshwater mammals. All that noise pollution and garbage thrown by humans interferes with their echolocation, while the dams that we build, fragments their habitat.

Add to that habitat loss, fishing gear that entangles them and also pollution, hunting... and you know why this species is endangered.

SOUND CHECK

When this dolphin breathes – in the same way as other dolphins, it surfaces to breathe – it emits a sound that's like, *shushuk, shushuk*. So that's why the species is also called Susu.

IF YOU WERE TO GIVE THE SPECIES A NEW NAME, WHAT WOULD YOU CALL THE DOLPHIN?

DID YOU KNOW?

The Ganges River Dolphin was named the 'National Aquatic Animal of India' in 2009.[3]

HELP THE DOLPHIN GET TO HER FISH

This dolphin is hungry, but there are so many objects in her path that are confusing her. Can you guide her to the fish, making sure she does not hit a propeller, get entangled in a net or ingest any plastic?

CAMP OUT WITH THE INDIAN TENT TURTLE

This mostly herbivorous reptile[4] can be found in India, Nepal and Bangladesh. Although a protected species, because of their prettiness, they are often captured for the pet trade.

Having turtles as pets is a terrible idea, as they belong in the wild, near river banks. Since your house is not a river, abandon the idea of asking for a turtle or any other wild animal as a pet.

HOW CAN YOU HELP?

To begin with, use less. Plastic for instance, often ends up in water, and causes distress to all sorts of marine life, even endangering their lives. Before you use another plastic cup for lime juice or buy a bottle of water while travelling, think of where it could land up once you're done using it.

In 2017, a video went viral across the world. It was of a sea turtle. It was not a happy one. A piece of plastic straw had got stuck in the turtle's nose! How awful is that? Imagine a 10 cm[5] straw wedged in your nose and you can do nothing to get it out.

So awful was the video, that across the world, and even in India, people started refusing to use plastic straws. They told restaurants, cafes and even coconut vendors to not use the same. A lot of these establishments now use paper or metal straws, and a lot of people realise that you don't actually need straws to drink milkshakes, lassis or orange juice. Of course, straws are crucial for people with disabilities, and so they are also useful, however not everyone needs them.

Make a list of all the single-use plastic that you don't need to use, and then, well, don't use it.

WHAT CAN YOU DO?

Straws are fun, specially to make guzzling noises right at the end of a shake and it even annoys adults. But they are usually made of plastic, just like takeaway forks, spoons, plates and containers. That's awful because it all becomes part of the 25,000 tonnes[6] of plastic our country dumps every day. Ditch the straw and make others ditch it as well. Come up with alternative ways to drink a thick shake or slurpee.

ICYMI

In case you are wondering, Ms Arati is the same Arati Kumar-Rao[7] mentioned in the chapter 'From the North-east, With Love'. She's documenting what she calls the 'slow violence of ecological degradation'. She uses photos, words and art to talk about landscapes, migration, biodiversity and climate change – and of course, the Gangetic Dolphin.

5. DEER, DACHIGAM

CHUNMUN TRAVELS TO KASHMIR'S DACHIGAM NATIONAL PARK TO PHOTOGRAPH A RARE DEER. SHE IS GUIDED BY MR NAZIR, ONE OF INDIA'S MOST EXPERIENCED FOREST RANGERS.

CHUNMUN, YOU MUST HAVE HEARD THE URDU COUPLET, 'IF THERE'S PARADISE ON EARTH, IT IS HERE IN KASHMIR.'

AH, YES! BY EMPEROR JAHANGIR, RIGHT?

YEP. NOW LISTEN TO ONE BY USTAD NAZIR HIMSELF!

IRSHAAD!

IF THERE IS PARADISE WITHIN KASHMIR, IT IS HERE IN DACHIGAM!

THESE SOUTH-FACING SUNLIT SLOPES ARE THE BEST SPOTS TO OBSERVE KASHMIR'S STATE ANIMAL, THE HANGUL!

MR NAZIR, I WAS TOLD THAT YOU KNOW THE FOREST LIKE THE BACK OF YOUR HAND... NOW, I COMPLETELY AGREE!

WE'RE IN LUCK TODAY, CHUNMUN. THE HANGUL IS OTHERWISE A VERY SHY CREATURE, QUICK TO DETECT HUMAN PRESENCE USING THE WIND AND ITS SENSE OF SMELL, AND TAKES COVER.

AND WHY WOULD IT NOT? THIS CRITICALLY ENDANGERED ANIMAL FACES A LOT OF PRESSURE FROM CATTLE-GRAZING, HUNTING AND THE INCREASING MILITARY ACTIVITY IN THE REGION. AFTER ALL, CONSERVING WILDLIFE IN REGIONS OF CONFLICT HAS ITS OWN CHALLENGES...

AAANRHH! AA NRHH!

LOOK, CHUNMUN, IT'S A HANGUL STAG OUR FOREST DEPARTMENT HAD RADIO-COLLARED LAST YEAR. HE'S A FULL-GROWN CHAP NOW!

AND HE'S IN RUT! HE'S APPROACHING THE HERD OF FEMALES, HOPING TO WOO A MATE BY BRAYING AND SHOWING OFF HIS GLORIOUS ANTLERS!

> **WHAT A FINE MODEL HE MAKES! THE WHITE RUMP CONTRASTS SO BEAUTIFULLY WITH HIS RUSTY-RED BODY!**

> **I HOPE SHE DOES FIND HIM AS ATTRACTIVE AS YOU DO, CHUNMUN... WE REALLY NEED MORE HANGULS IN DACHIGAM!**

There's something magical about Dachigam. You walk into a forest laced with snow-capped mountains, the towering deodars, pines and oaks clad in different colours depending on the season. In autumn, the forest turns into a riot of yellow and orange; and in winter it's a veritable Narnia. You can expect the Snow Queen to come in her finery, or even better the Hangul.

CHANGING FORESTS, SEASONAL CHANGES

As the season changes, we also adapt.
In winter, we snuggle into warm clothes, eat lots of soups and hearty meals, and go to bed earlier because the days are shorter. What else do you do?

FUN FACTS

1. The Hangul is a sub-species of the Elk, and is endemic to Kashmir and Himachal Pradesh.
2. A stag is said to be in rut when the breeding season for deer approaches. During this time, stags compete with each other to win female attention, by locking antlers, braying and decorating their antlers with foliage!
3. Dachigam translates to 'ten villages', in memory of the ten villages that were relocated when the area was declared a National Park.
4. Dachigam is also home to the Himalayan Black Bear.
Q. Another species of deer found in Dachigam, is known for yielding a perfume from its scent glands, and is hunted illegally to obtain this perfume. Can you name this species?

Stuff you look forward to as the season changes:

Season	Look forward to	Absolutely detest
Winter		
Spring		
Summer		
Monsoon		

But humans are not the only ones who adapt to seasons, animals do that too. For instance,

Bears hibernate in winter.
Squirrels gather food for the cold months.
Can you think of other birds, insects and animals who do this?

Even trees adapt to seasons. Have you seen how the place you live in is transformed every season? Think of winter, when humans add layers and trees shed leaves. And how when spring comes, trees burst into bloom.

BENGALURU'S AVENUE TREES

If you live in Bengaluru, or know a friend who does, then ask them how the city changes with the seasons. From December to June, the canopy is constantly ablaze with colours – whether it's the pink Tabebuias or the purple Jacarandas or the yellow Copper Pods. There's a super long list.

Started by Hyder and Tipu Ali,[1] tree plantation in Bangalore (now Bengaluru) grew substantially under the British rule, as horticulturists planted trees from across the world.[2]

Of course, if you're planting a tree, make sure you get indigenous species of trees. They live happier in the soil they belong in.

TRACK A TREE

What do we mean – track a tree? A tree can't move. How do you track something that's rooted to the ground, almost like ummm, forever? Alright, here's what you do, you track a tree by the season!

1. Go pick a tree, any tree – it could be in your school, outside your window, at home or in a park.
2. Give it a name – it could be Groot, it could be Anita or Subir, Potato or Ent, Raintree or Laburnum; just name your tree.
3. Now, find out everything you can about the tree. Tip: If you put up a photo of the tree on the iNaturalist app, it often tells you about the species. Got the species name? Excellent.
4. Next, make a season chart: Head to http://www.seasonwatch.in/ and download their chart.

READ

Try to walk around your neighbourhood with a guide by Karthikeyan S., called *Discover Avenue Trees*. The little pocket book comes packed with information on identifying trees, the time they flower and the birds that visit them.

BECOME A TREE HUGGER

Now that you've become an expert on trees in your area, it's time to take it up a notch. You need to figure out which tree is the most huggable – banyan, baobab, palm tree or coconut tree. Rank them in order of preference.

My Tree Hugging Chart

Tree Common Name	Tree Scientific Name	Huggable Factor 1–10, with 10 being most huggable	Comments
			Too skinny
			Could not put arm around the tree, too stout
			Just perfect for hugging, especially when having a bad day

DRAW

In four easy steps, learn how to draw the Hangul.

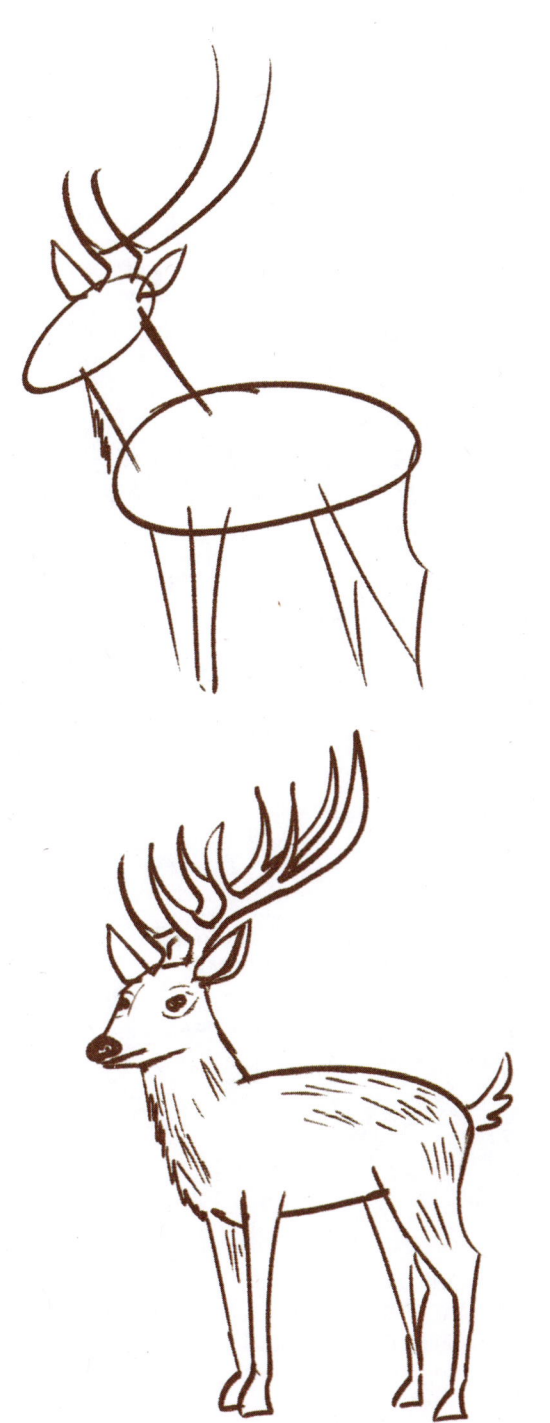

ICY MI

Nazir Malik is a forest ranger with Dachigam in Kashmir. He loves his park[3] so much that he's refused promotions so that he doesn't have to leave it. His favourite animal is the Himalayan Black Bear.

6 SPOTS IN THE FIELDS

CHUNMUN'S ASSIGNMENT TO PHOTOGRAPH LEOPARDS TAKES HER NOT TO A FOREST, BUT TO THE SUGARCANE FIELDS OF AKOLE, IN MAHARASHTRA. HERE SHE MEETS DR VIDYA, A CARNIVORE BIOLOGIST AND LEOPARD EXPERT.

I MAY HAVE TO DISAPPOINT YOU, CHUNMUN... YOUR CHANCES OF ACTUALLY SEEING A LEOPARD ARE RATHER BLEAK HERE!

BUT WHAT WE CERTAINLY WILL SEE, ARE SIGNS THAT LEOPARDS ARE THRIVING IN THESE FIELDS!

LEOPARDS SPECIALIZE IN STEALTH AND CAMOUFLAGE, AND THE SUGARCANE CROP OFFERS EXCELLENT COVER FOR THESE SECRETIVE CATS.

VIDYA, HOW IS IT THAT WHEN MAN-ANIMAL CONFLICT IS ON THE RISE THROUGHOUT THE COUNTRY, FARMERS HERE ARE CO-EXISTING WITH LEOPARDS IN RELATIVE HARMONY?

UNLIKE THE WEST, INDIA HAS TRADITIONALLY BEEN VERY TOLERANT OF WILDLIFE. IN FACT, TRIBES HERE WORSHIP THE LEOPARD GOD, 'WAGHOBA'!

MOST LEOPARDS HUNT STRAY DOGS HERE, BUT SOME DO KILL CATTLE. FARMERS HOWEVER, ARE FORBEARING, AS LONG AS THERE IS NO HARM TO HUMAN LIFE.

PEOPLE HERE HAVE ADAPTED TO LIVE ALONG-SIDE LEOPARDS FOR DECADES, JUST LIKE LEOPARDS HAVE ADAPTED FOR LIVING AMIDST PEOPLE...
SIMPLE MEASURES THAT THE VILLAGERS EMPLOY – WALKING WITH A TORCH IN THE DARK AND TALKING LOUDLY, HELP AVOID CONFRONTATIONS WITH THE CAT...
AND THIS ALSO TEACHES US SCIENTISTS A LOT ABOUT HOW PEOPLE AND WILDLIFE CAN COEXIST IN OUR COUNTRY!
AH, HERE WE ARE – A CAMERA TRAP I HAD SET UP LAST WEEK.
LET'S SEE WHAT THIS BABY'S CAUGHT.
WOW, VIDYA! IT'S A LEOPARDESS AND HER CUB!
WITH SOME EXPERIENCE, CHUNMUN, WE CAN IDENTIFY INDIVIDUAL LEOPARDS BY THEIR SPOTS... THIS ONE HERE IS A THIRD INDIVIDUAL, A MALE!
HERE'S A STRIPED HYENA STARING RIGHT INTO THE CAMERA TRAP! THESE SCAVENGERS FEED ON DEAD CATTLE CARCASSES AND OFTEN STEAL FROM LEOPARDS!
AND LOOK AT THIS! THE WORLD'S SMALLEST WILD CAT – THE RUSTY-SPOTTED CAT! IT'S CARRYING A KILL – A FIELD MOUSE!

A COOL CAT

Leopards are members of the cat family, just like lions and tigers. The spots you see on their body are called rosettes. You can figure why! They are shaped like a rose.

Leopards are:
- Nocturnal, they mostly hunt at night, and have cat naps during the day
- Solitary animals, they are loners
- Excellent climbers; after all, they are cats
- Found in India, China, Central Asia as well as sub-Saharan Africa, north-east Africa, Central Asia.

SPOT THE DIFFERENCE

Leopards are often confused with cheetahs. Which is rather sad as they are quite different.
Write all the differences you can see between these two images.

Leopard

Cheetah

A ROSETTE BY ANY OTHER NAME...

People visiting Kabini in Karnataka often come back with stories of having seen a black big cat. Those are black leopards that have excessive black pigment, but look carefully (at the photos) and you will see they also have rosettes on them.

TRAPPED!

Once when Dr Vidya Athreya checked the footage of the camera traps she had set up in Maharashtra, she found people imitating the leopard walk to be part of the camera trap images!

FUN FACTS

1. Camera traps are frequently used by wildlife scientists around the world to monitor animals. They capture images of wild animals passing by. Camera traps reveal the presence of rare and elusive creatures that may be extremely difficult to spot otherwise.
2. A leopard called 'Ajoba' that was radio-collared in Akole, was the subject of a Marathi feature film 'Ajoba'. The character played by Urmila Matondkar in that film was inspired by the real Dr Vidya (Vidya Athreya), who has been studying Akole's leopards.
3. Hyenas are extremely effective scavengers, and eat parts of carcasses that cannot be consumed by most other animals, such as hard bones!
4. A Russian sub-species of the Leopard is considered one of the rarest and most endangered cats in the world. Can you name this animal?

TRIVIA TODAY

Cheetahs once lived in India, but were hunted to extinction. Observe a minute of silence for extinct animals. However, now African cheetahs have recently been introduced to India, and conservationists are divided about whether this is a good idea or not. Time will tell.

COME UP WITH SOME ANIMAL PUN AND JOKES

Which animal is a rule breaker?
A cheetah (cheater)

WRITE AN APOLOGY LETTER TO A RODENT

In 2016, the world got some sad news. The Bramble Cay[1] melomys a.k.a. the Mosaic-tailed Rat was declared extinct because of climate change. They deserve an apology from us humans. Please write one. Make it scientific, add facts and figures, and be sincere in your apology.

....

....

....

....

....

WHO'S GONE THE DODO WAY? TICK THEM OFF!

Here's a list of some awesome species. Some have become extinct in the wild, some are holding on. Can you figure out which ones are no longer walking (or flying or crawling) on this Earth?

- Dodo
- Tasmanian Tiger
- Forest Owlet
- Passenger Pigeon
- Siberian Tiger
- Golden Toad
- Zanzibar Leopard
- Cheetah

Answers:
Dodo
Tasmanian Tiger
Passenger Pigeon
Golden Toad
Zanzibar Leopard

DRAW ONE OF THE EXTINCT ANIMAL SPECIES

DEBATE IT OUT

Congrats! You've made it to your school's debate team. You pull a topic out of the hat, and its 'India's decision to re-introduce the cheetah here'. Everyone knows that the cheetah went extinct in 1952; most of the world's cheetahs live in Africa. Now the government has brought them to Kuno National Park in Madhya Pradesh.

Debate the pros of this re-introduction using at least three arguments.
Now, debate the cons, again you get to use three arguments.

At the end of this, write down what you would do, if you were India's Environment Minister for a day.

ICYMI

Dr Vidya Athreya researches leopard ecology and human-leopard conflict. For her amazing work, she was honoured with the Carl Zeiss Wildlife Conservation Award.

READ THIS

Check out the book *10 Indian Champions Who Are Fighting to Save the Planet* by Bijal Vachharajani and Radha Rangarajan to read more about the amazing Dr Vidya Athreya.

7
THE FISHER FROM BENGAL
CHUNMUN'S SEARCH FOR THE ELUSIVE FISHING CAT TAKES HER TO THE SWAMPS OF HOWRAH IN WEST BENGAL. HERE SHE MEETS MISS TIASA, A CONSERVATION BIOLOGIST WHO STUDIES FISHING CATS.
OH, GOODNESS! THE SLUSH JUST GETS DEEPER AND DEEPER! TIASA, HATS OFF TO YOU FOR CONDUCTING YOUR RESEARCH HERE!
IT'S SLUSH FOR US. BUT FOR THE FISHING CAT, IT'S A PLAYGROUND!
HERE IN THE MARSHES OF THE LOWER GANGETIC FLOOD PLAINS, FISHING CATS HAVE TALL GRASS TO HIDE IN, AND AMPLE OPPORTUNITY TO HUNT FISH!
MY CAMERA TRAP TELLS ME THAT THIS POND IS VISITED BY A HUGE MALE EVERY NIGHT. LET'S WAIT HERE AND WATCH...
RROW-RROW-RROW-RROW-RROW!
DRAT! THERE ARE DOGS HERE... THERE GOES OUR CHANCE-
SHHHH! THAT'S NOT A DOG BARKING! IT'S A-
-CAT?!
SURPRISED?

BARKING ISN'T THE ONLY WAY IN WHICH THE FISHING CAT DIFFERS FROM OTHER CATS...
IT IS PERFECTLY ADAPTED FOR AN AQUATIC LIFE! ITS WATER-RESISTANT COAT, WEBBED FEET, RUDDER-LIKE TAIL ARE ALL REMARKABLE ADAPTATIONS TO SWIM AND CATCH FISH!
BUT THIS ALSO BRINGS THE FISHING CAT IN CONFLICT WITH FISHING COMMUNITIES RESIDING HERE...
THE SPECIES FACES A HUGE THREAT FROM URBANIZATION WHICH HAS CAUSED A RAPID DISAPPEARANCE OF WETLANDS AND MARSHES. TO ADD TO THIS, THE CATS WERE ALSO PERSECUTED BY FISHERMEN FOR COMPETITION OVER FISH, AND FOR KILLING LIVESTOCK.
BUT NOW, MY TEAM IS TRYING TO INVOLVE FISHING COMMUNITIES TO MONITOR FISHING CATS THAT VISIT THEIR PONDS, AND INVOLVE THEM IN CONSERVATION...
I GUESS THAT MUST'VE BEEN VERY EFFECTIVE IN GIVING NATIVE COMMUNITIES A SENSE OF OWNERSHIP OVER THEIR LAND...
AND ALSO A SENSE OF PRIDE OVER THE FISHING CAT...
AFTER ALL, THIS FISHERMAN IS THE STATE ANIMAL OF WEST BENGAL!
SPLASH!
CLICK!

> DID YOU GET THE SHOT YOU WANTED?

> YES! ONE OF INDIA'S RAREST CARNIVORES IN ACTION! I COULDN'T HAVE ASKED FOR MORE!

> WELL, I'M REALLY GLAD THAT BOTH OF YOU GOT YOUR CATCH TONIGHT!

IS THAT A DOG, SUPERMAN OR A FISHING CAT?

Isn't it just amazing the way the Fishing Cat sounds like a dog! So confusing for a dog, who must think, ooh look it's my friend, only to discover it's a cat.

Did you hear that? Duggu's barking in confusion just at the thought of this.

JUNGLE ORCHESTRA

Toucans croak like frogs
Hyenas laugh like humans
Parakeets and Parrots can mimic people
Racket-tailed Drongos can mimic the sounds of other animals and noises
Brahminy Kites wail like they are throwing a tantrum
Red-wattled Lapwings constantly ask, *did-you-do-it*?

Answers:

Fishing Cat - 3
Cheetah - 1
Rhino - 6
Wild Boar - 4
Giraffe - 5
Zebra - 2

CLOSE-UP CAMOUFLAGE

Here are some close-ups of animals. Can you guess from the pattern, which species it belongs to?

Fishing Cat .... Cheetah ....

Rhino .... Wild Boar ....

Giraffe .... Zebra ....

1

2

3

4

5

6

*Answers are on facing page.

POOPER SCOOPER

What's long and cylindrical in shape, and may have some white stuff on it?[1] Fishing Cat's poop.

If you're wondering why we are talking about poop, it's because this matter is essential for scientists to study wild animals. Called scat,[2] it tells them a lot about what the animal ate, what has it been up to and where all it has been roaming about. Also by taking the scat back to the lab, scientists can determine the animals' health by examining its DNA.

For instance, Uma Ramakrishnan[3] of the National Centre for Biological Sciences led a team to understand Similipal's black tigers. Yes, this national park in Odisha is home to tigers that are more black than orange. Their study showed that a single mutation in a gene causes the tiger stripes to become broader or spread across its body. They believe these mutations are occurring as Similipal's tigers live in isolation in the eastern part of India, and don't really move around. While species live in isolation, it places them at risk from life-threatening mutations as well as from disease outbreaks that can wipe out the entire population as they have nowhere to escape to.

THROUGH THE MICROSCOPE

You are a scientist and someone forgot to label all the poop in your lab! Can you figure out which animal's poop is which?

Fishing Cat ....

Tiger ....

Chinkara ....

Elephant ....

Bat ....

Answers:
1. Tiger
2. Fishing Cat
3. Elephant
4. Chinkara
5. Bat

FOLLOWING THE DUNG TRAIL

Inevitably, following dung in the jungle will lead you to... dung beetles!

These beetles are so important because they use poop to feed and nest in. Without them there'd be mountains of dung to wade through. No wonder the ancient Egyptians held the scarab beetles in high regard.

ICYMI

Tiasa Adhya co-founded The Fishing Cat Project to protect the species and its home, the wetlands. The cat needs wetlands for its food, and so Tiasa is making the connection between the species, its habitat and its importance to humans. She's even working with local people to monitor Fishing Cats in their area and name them.

FUN FACTS

1. Other than West Bengal, Fishing Cats have also been documented in Rajasthan, Uttar Pradesh and Andhra Pradesh. Coringa Wildlife Sanctuary in Andhra Pradesh has one of the largest populations of this cat.
2. Other than fish, Fishing Cats also prey on waterbirds, rodents, frogs, snakes, crabs and rarely, domestic cattle.
3. Other countries that the Fishing Cat is found in are Pakistan, Sri Lanka, Nepal, Bhutan, Bangladesh, Myanmar, Thailand and Cambodia.
Q. The Fishing Cat is the state animal of West Bengal. Can you name its state bird? (Hint: it also has something to do with fish!)

8
WARBLER AT THE WINDOW

ON HER BREAK, CHUNMUN VISITS UNCLE BIKKY AND DUGGU, AND MAKES AN ACQUAINTANCE WITH A CHIRPY WINTER VISITOR.

WOW, CHUNMUN! I ENVY YOU SO MUCH LOOKING AT THESE PICTURES!

ALL THESE FABULOUS CREATURES YOU'VE BEEN SHOOTING, AND ALL THE WONDERFUL PEOPLE YOU'VE BEEN MEETING AND LEARNING FROM!

I MISS YOU IN THE FIELD, UNCLE BIKKY! PLEASE FIX YOUR KNEES SOON, OLD MAN!

I MISS THE FIELD TOO, CHUNMUN... BUT DUGGU'S BEEN GIVING ME GREAT COMPANY, AND WE ALSO HAVE A FRIEND FROM THE HIMALAYAS VISITING US...

48

A FRIEND FROM THE HIMALAYAS?! DO YOU MEAN THAT DOCUMENTARY FILMMAKER FROM HIMACHAL WITH WHOM YOU HAD AN AFF-
SHHH!!
SHE'S CALLING!
CHI-LIP!
THERE SHE IS- THE GREENISH WARBLER!
CHI-LIP!
IMAGINE – THIS TINY BIRD THAT COULD FIT INTO YOUR PALM, CROSSES THE MIGHTY HIMALAYAS EVERY WINTER TO VISIT OUR BACKYARD!
CHI-LIP!
HER CHIRPING WHICH DELIGHTS OUR EARS, IS ACTUALLY A WARNING FOR OTHER WARBLERS TO KEEP AWAY FROM HER TERRITORY!
CHI-LIP!
THIS RESTLESS BIRD FLITS ABOUT FROM LEAF TO LEAF ALL DAY, HUNTING INSECTS AND THEIR LARVAE! NO WONDER THIS GENUS OF BIRDS IS CALLED 'LEAF WARBLERS'...
CLICK-CLICK-CLICK!
HEY, I THOUGHT YOU'RE ON A BREAK, CHUNMUN!
ONCE A PHOTOGRAPHER, ALWAYS A PHOTOGRAPHER!

> WELL, DUGGU, THE WARBLER ISN'T THE ONLY RESTLESS VISITOR TO OUR HOUSE TODAY!

NOT A RARE WINDOW

Who'd have thought the world would talk so much about windows. Still, that's what happened in 2020 and 2021, when the COVID-19 pandemic struck. At that time, many people got a renewed appreciation for views from their window, because it was that slice of rectangle/square/circle/triangle that was literally a window into the world outside.

Look out of your window now, and make a list of all the things you see. Make it as detailed as you can. Include fluffy clouds and wisps of clouds, pigeons and crows, wagtails, swifts and other birds, trees, buildings, uncles, aunties, flapping clotheslines, houses… include everyone and everything you see.

VIEW FINDER

Grab a sheet of paper and draw the view you can see. Remember, you don't have to be a picture-perfect artist. You can make cartoons, draw abstract lines, make stick figures or create a collage.

FUN FACTS

1. Warblers are among the most difficult birds to tell apart! Many species look similar to each other and only bear minute differences, making them a puzzle for birdwatchers.
2. Warblers may look tiny and drab, but they are champions of migration! If you happen to spot one in your backyard in India, chances are that it has flown thousands of kilometres from the northern latitudes to visit you!
3. Warblers are also accomplished vocalists. A species of warbler, the Marsh Warbler, can mimic up to 80 species of birds!
Q. A backyard warbler commonly seen in northern India, that appears quite similar to the Greenish Warbler, is named after an eminent British ornithologist who was also one of the founders of the Indian National Congress. Can you name this warbler?

WHAT'S IN A NAME?

Someone clearly ran out of name ideas when naming the Greenish Warbler. The other name that this species of migratory bird is known by is Dull Green Warbler. Whaaaat! You can definitely come up with better names for this bird that travels so many miles. Something NOT dull!

Here are some clues about the bird that would help you name the species:
- The warbler is greyish-green[1] in colour.
- It has a white eyeliner and a white stripe above its eye.
- Their call is really distinctive, piercing and trilling.

DID YOU KNOW?

Remember how the British had winter homes and summer homes in India? The Greenish Warbler too flits between its summer home in Eurasia and the high Himalaya and its winter home in the rest of India. There's also the Red-breasted Flycatcher,[2] Common Rosefinch and the Bluethroat, all of which like to stay here in winter.

POLLINATION PROS

Warblers are also pollinators; they can transport pollen grains to and fro. Without pollinators, we will have no food to eat – and that's a fact! In addition to birds, bees are pollinators too. More information on that coming up below!

STATE IT HERE

Find out the official animal of your state. Yes, we have state animals too. Write it down here.

LIST TIME!

Grab your pens, your help is needed urgently! Here's a flock of the official birds of India's states and union territories.[3] But they all seem to have got lost thanks to some electric wire lines, road construction, disappearing forests and habitat loss. Can you help them find their way to their home state? Hint: Some birds belong to multiple states.

Place the serial number of the state or union territories against its state bird

Answers are on page 59

- 6. Jammu and Kashmir
- 7. Ladakh
- 9. Himachal Pradesh
- 2. Chandigarh
- 20. Punjab
- 26. Uttarakhand
- 8. Haryana
- 8. Delhi
- 22. Sikkim
- 2. Arunachal Pradesh
- 3. Assam
- 18. Nagaland
- 16. Meghalaya
- 15. Manipur
- 21. Rajasthan
- 27. Uttar Pradesh
- 4. Bihar
- 25. Tripura
- 17. Mizoram
- 7. Gujarat
- 13. Madhya Pradesh
- 10. Jharkhand
- 28. West Bengal
- 5. Chhattisgarh
- 19. Orissa
- 14. Maharashtra
- 24. Telangana
- 6. Goa
- 1. Andhra Pradesh
- 11. Karnataka
- 12. Kerala
- 23. Tamil Nadu
- 5. Puducherry
- 4. Lakshadweep
- 1. Andaman & Nicobar Islands

○ Indian Bustard
○ Indian Paradise Flycatcher
○ Blood Pheasant
○ Flame-throated Bulbul
○ Western Tragopan
○ Indian Roller
○ Himalayan Monal
○ Asian Koel

○ Hill Myna
○ Black Francolin
○ Peacock
○ Hill Myna
○ Black-necked Crane
○ Blyth's Tragopan
○ Great Hornbill

○ Sarus Crane
○ Blood Pheasant
○ White-winged Wood Duck
○ Imperial Pigeon
○ Greater Flamingo
○ House Sparrow
○ Koel

○ Green Imperial Pigeon
○ Mrs. Hume's Pheasant
○ White-throated Kingfisher
○ Andaman Wood Pigeon
○ Northern Goshawk
○ Emerald Dove
○ Sooty Tern

FIND A BEE-PHOBIC FRIEND WHO THINKS BEES BITE AND NOT STING, AND EXPLAIN THE CONTRARY TO THEM.

Not only do bees not want to sting you, they're also pollinators. We need them to be able to eat! But alas, pesticides are taking a toll on bee populations. In fact, it's so bad, that authors are writing dystopian books set in the future where they imagine a world with no bees.

DRAW

So weird that people seem to start yelling at the top of their voices when they spot a bee close to them. Bees are really not trying to sting you... calm down. In fact, they are the ones under threat, from pesticides to monoculture agricultural practices. This despite there being 20,000 species of bees[4] in the world. Learn how to draw a pollen-carrying bee in three easy steps.

SPARROW SPECIAL

Real estate in cities is expensive, ask the grown-ups, they will agree. While pigeons and crows seem to have figured out their living situations in cities, a species that's found it tough is the sparrow. They can't figure where to nest, nor can they easily find insect larvae for their fledglings. But you can help. Build your own sparrow home and invite them to come and nest.

There are plenty of tutorials online, but it's basically a wooden box with a hole big enough for a sparrow to flit in and out of, but small enough for predators to stay away.

If you decide to build one, then make sure you keep the box in a place that's safe and free from interference and predators. And you need to leave it alone. Observe from a respectful distance as the Common House Sparrow becomes your common house guest.

9

RAVINE REPTILIAN

CHUNMUN'S QUEST TO PHOTOGRAPH A CRITICALLY ENDANGERED CROCODILE TAKES HER TO THE RAVINES OF THE CHAMBAL RIVER IN RAJASTHAN. HERE SHE IS GUIDED BY CONSERVATION BIOLOGIST MR DHARAM.

LOOK, CHUNMUN — AN INDIAN SKIMMER SKIMMING THE WATER SURFACE FOR FISH WITH ITS ELONGATED LOWER BILL!

AREN'T SKIMMERS INDICATORS OF A HEALTHY AQUATIC ECOSYSTEM, MR DHARAM?

THAT'S RIGHT, CHUNMUN! AND SO IS OUR TARGET SPECIES, THE GHARIAL.

SCREECH

DO YOU SEE THAT BULGE IN THE WATER?

THAT LITTLE ROCK?

HAHAHA, YES... THAT 'LITTLE ROCK' IS THE REASON WHY A GHARIAL IS CALLED A GHARIAL! IT'S THE 'GHARA' OR 'POT' ON THE MALE'S SNOUT!
WOW! FOOLED ME!
MALE GHARIALS USE THESE POT-LIKE PROJECTIONS TO MAKE BUBBLES AND BELLOWING SOUNDS, TO ATTRACT MATES AND THREATEN RIVALS...
GHARIALS TOO, LIKE SKIMMERS, ARE FISH-EATERS, AND THRIVE ONLY IN UNPOLLUTED WATERS...
AND CHAMBAL HAS SOME OF THE MOST PRISTINE AND UNTOUCHED WATERS IN OUR COUNTRY, THANKS TO AN OLD MYTH THAT THE RIVER IS CURSED!
I'M SURE THIS CURSE HAS BEEN A BOON IN DISGUISE FOR CHAMBAL'S WILD RESIDENTS...
THAT'S RIGHT, CHUNMUN, BUT NOT FOR LONG... POPULATIONS OF ALL WILD CREATURES IN CHAMBAL ARE FACING GROWING THREATS AND STEEP POPULATION DECLINES.

THE GHARIAL IS INCREASINGLY THREATENED BY DAMS, WATER POLLUTION AND ENTANGLEMENT IN FISHING NETS.

THOSE SAND-MINING TRUCKS YOU SEE THERE, ARE THE GHARIAL'S WORST ENEMIES! THAT IS BECAUSE GHARIALS NEST IN THE CHAMBAL'S SANDY BANKS...

AN ANIMAL THAT HAS REMAINED UNCHANGED SINCE DINOSAURS ROAMED THE EARTH, IS NOW FACING THE THREAT OF EXTINCTION... WHAT A PITY IT WOULD BE IF WE ARE UNABLE TO MAKE GHARIALS BOUNCE BACK!

POP! BLGGGGRRHHH! POP! POP! BLGGGGRRHHH! POP!

OUR FRIEND IS BELLOWING IN COMPLETE AGREEMENT!

HAHAHA!

If you thought that the Gharial got its name from the Gujarati word for clock, you're wrong. The name comes from *ghara*, like the pot. Because the male reptile has a knob at the end of its snout.

FUN FACTS

1. The Chambal is a tributary of the Yamuna. It originates in the Janapav Hills of the Vindhya Range, and flows through three states – Madhya Pradesh, Rajasthan and Uttar Pradesh – before draining into the Yamuna.
2. The Chambal Basin is home to several species of freshwater turtles. Among these is the critically endangered Red-crowned Roof Turtle.
3. Skimmers are the only birds with an uneven bill arrangement, with the lower bill longer than the upper. They use this apparatus to skim the water, snapping it shut when a fish is caught. This feature gives them the nickname 'scissorbill'.
4. Unlike most reptiles, Gharials are devoted fathers! Hundreds of Gharial hatchlings can be seen basking and riding on the father's back during the breeding season.
Q. The Chambal Basin is home to another species of crocodile, much more common than the Gharial. Can you name this species?

GUESS WHAT?

If you've seen Gharials basking in the sun, you'd think they never move. But studies have shown that they actually travel long distances during the monsoon for feeding, and after that to bask and lay their eggs.

TIME'S TICKING FOR THE GHARIAL

You already know from the comic that the Gharial is a critically endangered species. What's happening is that these reptiles love to chill in fresh water – rivers that are clean[1] and fast-flowing – but projects such as dams make their rivers into lakes, which they don't like. And even sand removal and mining impacts them, by forcing them to leave the area or destroying their nesting site. On top of that, because they have long snouts, they often get entangled in fishing nets.

FIRST, OF HOPEFULLY MANY

Guess what? Chambal is India's first riverine sanctuary. Apart from the Gharial and the Red-crowned Roofed Turtle, the National Chambal Gharial Wildlife Sanctuary[2] is also home to the Gangetic Dolphin, Sarus Crane, Indian Skimmer and Pallas's Fish Eagle.

TURTLE, SPLURTLE

The Chambal river is home to Gharials, yes. But it's also home to a species that is only found in South Asia – the Red-crowned Roofed Turtle. This is really sad, because earlier the turtle was found across the Ganges river[3] in both India and Bangladesh.

GET THE TURTLE TO THE RIVER

The Red-crowned Roofed Turtle hatchling has just come out of her shell. Would you help guide her to the river, so that she can swim and make a new life for herself?

TALL AS A CRANE

The next time you want to describe yourself as very tall, we've got a simile for you. You can say, I am as tall as a Sarus Crane. That's because it's the tallest flying bird on the planet! They are the same height as a person who is 5 feet tall, and their wingspan is 7 feet 8 inches! That makes them taller than Amitabh Bachchan, by 1 foot 7 inches.

ICYMI

Conservationist Dr Dharmendra Khandal (whom Mr Dharam is based on), works in Chambal and Ranthambhore in Rajasthan. He has been speaking out for protecting the riverine sanctuary.

READ

Check out Aparna Kapur and Rosh's *Ghum-Ghum Gharial's Glorious Adventure* that tells the story of a baby Gharial who gets lost and needs to find her way home.

Answers for 'List Time' on page 52

State Birds
1. Andhra Pradesh – Indian Roller
2. Arunachal Pradesh – Great Hornbill
3. Assam – White-winged Wood Duck
4. Bihar – Indian Roller
5. Chhattisgarh – Hill Myna
6. Goa – Flame-throated Bulbul
7. Gujarat – Greater Flamingo
8. Haryana – Black Francolin
9. Himachal Pradesh – Western Tragopan
10. Jharkhand – Asian Koel
11. Karnataka – Indian Roller
12. Kerala – Great Hornbill
13. Madhya Pradesh – Indian Paradise-Flycatcher
14. Maharashtra – Yellow-footed Green Pigeon
15. Manipur – Mrs Hume's Pheasant
16. Meghalaya – Hill Myna
17. Mizoram – Mrs Hume's Pheasant
18. Nagaland – Blyth's Tragopan
19. Odisha – Indian Roller
20. Punjab – Northern Goshawk
21. Rajasthan – Indian Bustard
22. Sikkim – Blood Pheasant
23. Tamil Nadu – Emerald Dove
24. Telangana – Indian Roller
25. Tripura – Green Imperial Pigeon
26. Uttarakhand – Himalayan Monal
27. Uttar Pradesh – Sarus Crane
28. West Bengal – White-throated Kingfisher

Official Birds of Union Territories
1. Andaman & Nicobar Islands – Andaman Wood Pigeon
2. Chandigarh – Indian Grey Hornbill
4. Lakshadweep – Sooty Tern
5. Puducherry – Asian Koel
6. Jammu and Kashmir – Black-necked Crane
7. Ladakh – Black-necked Crane
8. Delhi – House Sparrow

10
The Humpback of the Arabian Sea

Earlier, Chunmun photographed a freshwater dolphin. Now it's time for a marine one — the Indian humpback dolphin! Chunmun visits Goa and meets Puja, a dolphin tour operator and conservationist who guides her on this task.

All set, team? 3... 2... 1...

VROOOM!

Here we go!

Photographing dolphins from a moving boat on these rocky waves is going to be tricky!

Don't worry, Chunmun. We shut our engines the moment we spot a dolphin pod. We're an ethical dolphin-watching tour company!

Dolphins of the west coast face a lot of threats. Many get entangled in fishing nets, and the rising pollution makes the marine environment toxic for their survival...

60

TOURISM ADDS ITS OWN SHARE OF PRESSURES. DOLPHINS USE ECHOLOCATION TO COMMUNICATE. NOISE FROM VESSELS IS A HUGE CAUSE OF DISTURBANCE TO DOLPHIN ECOLOGY, AS IT HAMPERS THEIR COMMUNICATION UNDERWATER.

BUT ON OUR TOURS, WE ENSURE THAT ETHICAL DOLPHIN-WATCHING PROTOCOL IS STRICTLY FOLLOWED. OUR TEAM OF LOCAL BOATMEN BRING IN TONNES OF EXPERIENCE IN TRACKING DOLPHINS, AND THIS BECOMES A BENEFICIAL PARTNERSHIP FOR BOTH!

DID YOU SEE THAT?! A PINK FLIPPER JUST SURFACED!

MISTER MORJE, COULD YOU SWITCH THE ENGINE OFF? THERE ARE DOLPHINS AROUND.

REALLY?!

RIGHT AWAY, PUJA!

THERE! IT'S A MOTHER WITH HER CALF!

WOW, I CAN CLEARLY SEE THE 'HUMP' ON THE DORSAL FIN, AFTER WHICH THE INDIAN HUMPBACK DOLPHIN GETS ITS NAME!

PRECISELY, CHUNMUN!

ETHICAL TOURISM

Choose Your Adventure Trip

It's that time of the year. School's finally over, vacations have started.
You are standing by:

1. Forest

2. Marine site

3. Grassland

FUN FACTS

1. India has 16 species of dolphins! Of these, the Gangetic Dolphin is the only freshwater dolphin.
2. Young Indian Humpback Dolphins are grey in colour. As they mature, they develop pink speckles.
3. Indian Humpback Dolphins can grow upto 9 feet!
4. A mass of adipose tissue present in the heads of dolphins and whales acts as a sound lens and helps in echolocation. This tissue is commonly referred to as the 'melon'.
Q. Goa's coast is also home to the world's most familiar dolphin, also the largest species of beaked dolphin. Can you name this animal?

CHOICE #1
Forest

You put on sunscreen, get a broad-rimmed sun hat and climb into the canter. Your canter gets route number 4. You all drive in, and the guide finds out that there's a sighting of a tiger on route number 7. Everyone says they want to go there, though canters are expected to stick by their route.
What is your response?

Yes – My parents paid a lot of money for this trip, let's go to 7. Go to **1. Y.**
No – Let's stick to our route. Go to **1. N.**

- **1. N.** Well done, the forest is not only about seeing one charismatic species. Sit back and take in the ecosystem. Spot different birds – is that an owlet in a tree hole and a crested serpent eagle circling above you? Keep an eye out for sloth bears, mongooses, peacocks, crocodiles, leopards and of course cat poop.
- **1. Y.** Boo, wrong choice. Routes are given out to canters so as not to overcrowd the forest trails and stress the animals out.

Your guide spots a tiger! YAY! But no one can see the cool cat basking behind the bush because she's so hard to spot! You...
- Start yelling and suggest the canter go off track to get a closer look at the cat. One can only crane their neck this much. Go to **1.1. Y.**
- Wait patiently for the tiger to move. Go to **1.1. N.**

- **1. 1. Y.** Eye roll... that's a terrible idea. Animals do not like this. Imagine someone rushing at you just to look at you closely and yelling at you especially when you have super sensitive hearing. So creepy.
- **1. 1. N.** Good choice, now wait and we're sure you will meet the tiger. And, even if you don't, its all good.

CHOICE #2
Marine Site

Your life jacket's strapped on, you're on a boat and ready to go for a dive or to watch dolphins. You have already switched places with the person threatening to be seasick and are now at a safe distance from projectile vomit. Suddenly a fin appears.

- You urge the boat to go closer, a fin's just not enough. Go to **2. Y.**
- You catch your breath and wait for the mammal to show itself. Go to **2. N.**

2. Y. The dolphin's already stressed, thanks to noise pollution and habitat disturbances. Nosy loud boats are so not welcome.

2. N. Aha, good things come to those who wait. There, the dolphin's smiling at you.

CHOICE #3
Grassland

You've got your binoculars that are longer than Pinocchio's nose after multiple lies, and have woken up at the crack of dawn to catch the early bird who is out to get the worm. After several frustrating minutes of not spotting that one bird you came to spot, your friend removes his phone and says, I have an idea. I will play the bird's call and lure it here. Your reply:

- YEAH, let's do that, do you need me to press play? Go to **3. Y.**
- Dude, I can't believe you said that. Go to **3. N.**

3. Y. Hope your friend's battery drains out just for that terrible behaviour. Misusing playback is illegal in parts of the world because it's harmful and can stress the bird. It can confuse them[1] and even disturb their social behaviour.

3. N. You're a good egg, now go find another bird. There's plenty in the bush. No one needs them in the hand.

ICYMI

With a master's degree in biodiversity conservation, Puja Mitra started Terra Conscious in Goa to offer ethical dolphin watching tours that don't stress the mammals.[2]

DRAW

Figure out how to draw the Indian Humpback Dolphin and her calf. Don't forget the pink speckles for the mamma.

11

WHERE FROGS FLY!

CHUNMUN ACCOMPANIES HERPETOLOGIST DR BIJU ON A RESEARCH EXPEDITION TO THE WESTERN GHATS IN KERALA, IN PURSUIT OF FROGS.

YOUR CAMERA'S READY, RIGHT, CHUNMUN? BECAUSE THIS FROG IS ABOUT TO DO SOMETHING TRULY BIZARRE!

WOW! WHAT IS THIS MOVE?!

SAY HELLO TO THE INDIAN DANCING FROG! TO ATTRACT A MATE, THIS FROG EXTENDS ITS HIND LEG TO SHOW OFF ITS FLASHY WEBS!

LOOK UP THERE!

WAIT, AM I REALLY SEEING A FROG FLY?!

WELL, NOT EXACTLY FLY, BUT GLIDE. THIS MALABAR GLIDING FROG HAS LARGE PADS ON ITS TOE WEBS, THAT CAN BE DEPLOYED LIKE A PARACHUTE TO CROSS STREAMS OR ESCAPE PREDATORS!

CHECK OUT THOSE FUNKY PATTERNS ON THE BODY OF THIS FUNGOID FROG...
BEATS A TIGER'S STRIPES OR A LEOPARD'S SPOTS IN STYLE ANY DAY!
THIS PIG-LIKE FROG, THE INDIAN PURPLE FROG, LIVES ITS ENTIRE LIFE UNDER-GROUND, EMERGING ONLY IN THE MONSOONS TO BREED!
WOULD YOU BELIEVE THIS CHUNMUN? ALL THE SPECIES WE'VE ENCOUNTERED TODAY ARE ENDEMIC TO THE WESTERN GHATS, AND FOUND NOWHERE ELSE ON EARTH!
INCREDIBLE! BUT WOULDN'T THIS ALSO MEAN THAT THEY'RE ALL SEVERELY THREATENED?
YOU'RE RIGHT. IMAGINE WHAT CLIMATE CHANGE AND UNTIMELY MONSOONS CAN DO TO A CREATURE WHOSE LIFE REVOLVES AROUND RAINS!
AS FROGS BREATHE THROUGH THEIR SKIN, THEY'RE THE FIRST TO BE AFFECTED BY CHANGES IN THE ENVIRONMENT...
THIS IS WHY THEY'RE CALLED BIOINDICATORS.
HERE IN THE WESTERN GHATS, HABITAT LOSS, DEFORESTATION, DAMS AND MINES MAKE IT EVEN WORSE FOR FROGS!

> FROGS DESERVE SO MUCH BETTER THAN THIS, DR BIJU!
>
> YOU KNOW, I'VE NOT HAD A SINGLE MOSQUITO BITE TONIGHT...
>
> AND I'M QUITE SURE THAT IT'S NOT JUST MY MOSQUITO-REPELLANT AT WORK!

If you thought you were going to see frogs at ground level only, think again! We are looking at frogs that fly, rather glide, like the Malabar Gliding Frog that's found in the Western Ghats.

If you've ever seen someone hang-gliding or parachuting, you will realise that the wind does a lot of the work. Which is exactly what gliding frogs tap into. They use the webbing between their fingers and toes to parachute from the top of trees. Are you watching, Spiderman? Watch and learn the next time you're facing Green Goblin or one of your many nemeses.

Rule Of Thumb

The Small Gliding Frog (*Rhacophorus lateralis*) is one of the smallest gliding frogs, measuring just 4 cms.[1] That's about as big as your parent's thumb. The species was thought to have disappeared for a long time, and was then rediscovered in the Madikeri forests in the late nineties.

FUN FACTS

1. Over 90 species of frogs found in the Western Ghats are endemic to the region!
2. Another species of frog, the Anamalai Gliding Frog, found in the Anamalai Hills in Tamil Nadu and Kerala (a part of the Western Ghats), can glide like the Malabar Gliding Frog.
3. The Ochlandra Bush Frog found in the Western Ghats resides in the stems of Ochlandra, a variety of bamboo!
4. The Indian Purple Frog's Latin name, *Nasikabatrachus sahyadrensis* is derived from two Indian words: 'nasika' meaning 'nose' in Sanskrit, and 'Sahyadri', the local name for the Western Ghats.
Q. A group of limbless, snake-like amphibians are also found in the Western Ghats, many species of which are endemic to the region. Can you name this group of amphibians?

LEAP TO A VERSE

We've taken a stab at some froggy poetry here. After that it's your turn. To be sung to Don McLean's *Starry, Starry Night*.

Starry, starry frog
Hiding under a rotten log
Your silence thick, like a fog
Without a mating call,
how do you attract another frog?

Oh dear frog in the Western Ghats
Because of our belching GHGS, methane farts
And pollutants that industries spew
Now climate change is threatening you
When I look at your arrow head
My heart fills with a lot of dread
Because you breathe through your skin
Now your life's going down the bin

Purple frog, purple frog
Purple frog, purple frog
Purple frog, purple frog
I only wanted to see you, purple frog
But you never come out, purple frog

If I had a dog, I'd name her Frog
We'd live in a bog, and sleep on a log.
She'd growl, and I would croak
Grr, Trr, Grr, Trr.
Can you hear us under the mighty oak?

**Think you can rhyme better?
Here's your chance.**

ICYMI

Dubbed the Frogman of India, Biju S.D. has been studying amphibians for many decades. Along with his students, he's discovered as many as 100 amphibians in India[2] and Sri Lanka. As part of his work, he tries to find species that are considered lost.

TOGETHER!

A group of animals are often defined by the use of a collective noun. What's really cool about frogs is that they have quite a few collective nouns.

They're generally called an army of frogs or a knot, but they are also called a band or a cohort. But when a group of male frogs is croaking away for a mate, they are called... a chorus!

CRACK THE ULTIMATE COLLECTIVE NOUN QUIZ

Collective nouns are used to describe a group. Do you know the names of these groups? Ummm, do not search for the answers online.

1. Owls
2. Crows
3. Lady bugs
4. Hedgehogs
5. Hippopotamus
6. Jellyfish
7. Hornets
8. Lizards
9. Cobras

1. Parliament
2. Murder
3. Loveliness
4. Array
5. Bloat
6. Bloop/Smack
7. Bike/Nest
8. Lounge
9. Quiver

IDENTIFY THESE ANIMALS BY THEIR MARKINGS

Here are some extreme close-ups of animals and insects, can you guess who these patterns belong to?

Galaxy Frog ....

Crocodile ....

Tiger Moth ....

Jewel Beetle ....

Cheetah ....

Answers:
1. Cheetah
2. Jewel Beetle
3. Tiger moth
4. Galaxy Frog
5. Crocodile

71

12

Macro Magic!

CHUNMUN MEETS UNCLE BIKKY'S FRIEND AND NATURALIST MR KARTHIKEYAN, WHO OPENS UP A WHOLE NEW WORLD FOR HER TO EXPLORE – THAT OF INSECTS.

AH, THERE YOU ARE, MR. KARTHIKEYAN! UNCLE BIKKY SAID THAT THE BEST WAY TO IDENTIFY YOU IS TO LOOK FOR A MAN IN A HAT, TURNING STONES OVER TO PHOTOGRAPH INSECTS!

HAHAHA! I CAN'T DENY THAT! PLEASED TO MEET YOU, CHUNMUN!

I KNOW YOU'VE BEEN DOCUMENTING SOME OF THE RAREST BIRDS AND MAMMALS, BUT ONCE YOU START OBSERVING INSECTS, YOUR ENTIRE VIEW OF NATURE GETS AN UPGRADE!

TAKE A LOOK AT THIS TWIG FOR EXAMPLE...

PEEP CLOSER AND YOU'LL NOTICE THAT THE TWIG IS ACTUALLY A CAMOUFLAGED OWLFLY! CALLED SO BECAUSE OF ITS LARGE, OWL-LIKE EYES, THIS INSECT PERCHES WITH ITS STICK-LIKE ABDOMEN HELD PERPENDICULAR TO THE BRANCH, RESEMBLING A TWIG!

72

CAN YOU GUESS WHAT THOSE PITS IN THE GROUND ARE?
UMM... EARTHWORM TUNNELS?
NO. THESE ARE TRAPS SET FOR ANTS BY A VERY CLEVER PREDATOR – THE LARVA OF AN ANTLION!
THE PITS ARE DESIGNED WITH SUCH MATHEMATICAL PRECISION, THAT ONCE AN ANT SLIPS INTO IT, THERE'S NO WAY FOR IT TO CLIMB OUT, AND IT FALLS STRAIGHT INTO THE JAWS OF THE ANTLION GRUB!
SPEAKING OF ANTS, WHAT KIND OF ANT IS THIS ONE HERE?
AND WHY IS IT SOLITARY?
GOOD QUESTION, CHUNMUN. BUT YOU SEE, THIS ISN'T AN ANT AT ALL!
IT IS AN ANT-MIMIC SPIDER, THAT PRETENDS TO BE AN ANT, AND PREYS ON ANTS!
ASTOUNDING! I NOT ONLY HAVE TO LEARN TO SHOOT MACRO, BUT ALSO SEE MACRO!
THE BEAUTY OF OBSERVING INSECTS IS THAT THEY'RE EVERYWHERE! WHO COULD GUESS THAT WE'VE SEEN SO MUCH WILDLIFE IN BANGALORE'S CUBBON PARK?

> **THE WORLD OF INSECTS IS JUST LIKE AN ANTLION'S PIT, YOU KNOW!**
>
> **ONCE YOU'RE IN IT, YOU CAN NEVER COME OUT?**
>
> **HAHA, EXACTLY!**
>
> **AND THE PLUS IS THAT WATCHING INSECTS MAKES YOUR EYES REALLY SHARP... YOU WON'T NEED SPECS EVEN AT MY AGE, UNLIKE YOUR ORNITHOLOGIST UNCLE!**

An Owlfly and an antlion may look just like a dragonfly. But don't be fooled. Karthikeyan S.[1] says that if you look at the owlfly closely, you will notice how different it is by simply looking at its antennae. It has a knobbed one, like you can see in the comic. Dragonflies have short antennae that look like wires.

Both of these are netwinged insects[2] that are part of an order called *Neuroptera*, whose members feed on insects that can be harmful to other trees and plants. Which means, they are rather useful.

OWLFLY

Knock, knock
Who's there?
Owlfly
Where will you fly?
Oho, owlfly!
Yes, where will you fly? Hello? Looks like she finally flew away.
Wonder where she went.

FUN FACTS

1. There are more species of insects in the world than there are any other species of animals! Around 90 per cent of all life forms on earth are insects!
2. The antlion larva is not just a predator par excellence, but a math genius! It builds its ant traps at an exact 45 degree angulation. When an ant slips into it and struggles to climb out, the antlion larva shoots sand at the ant to make it slip even faster!
3. Owlflies can be told apart from dragonflies by their knobbed antennae.
4. While ant mimic spiders mimic ants to hunt them, other insects may mimic ants for defence, because their predators avoid ants as food! Nymphs of certain species of crickets and grasshoppers mimic ants for defence.
Q. Can you think of other examples of insects mimicking other insects?

WRITE A SHORT STORY

Antlions are called thus because their larva is like a predator! The larva traps ants and other tiny insects to feed the adult antlions.

Have you read Rudyard Kipling's *Just So Stories*? Written way back in 1902, these stories were the author's way of coming up with innovative stories on how animals got their distinctive characteristics. Like how the leopard got its spots or the camel a hump. Of course, there are scientific reasons behind them, but he wove in the fantastical to spin tales about them.

Now it's your turn. Use your imagination to come up with a story, a fabulous, fantabulous, fantastical one about how in the world did the antlion get its name.

DRAGONFLY OR OWLFLY?

Below are two drawings of winged insects. Figure out which one's the dragonfly and which one's the owlfly.

DID YOU KNOW?

The dragonfly is also called the 'darner', 'devil's darning needle' or the 'devil's arrow'.
What is its name in your language?

ALSO!

Dragonflies are awesome because these shimmering beauties love to slurp up mosquitoes.

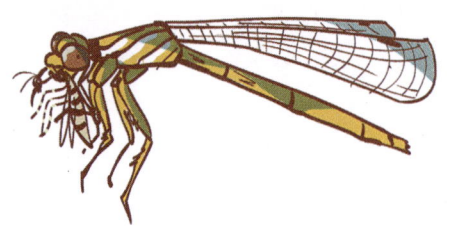

FIND OUT WHICH SNAKES ARE POISONOUS

Answer: None! This is a trick question as snakes are venomous, not poisonous. Ha!

The majority of Indian snakes are non-venomous. The Big Four that everyone keeps referring to, you should also know. Those are the Common Cobra, Common Krait, Russell's Viper and the Saw-scaled Viper. Go find out the scientific names for these snakes.

ET IN YOUR BACKYARD

Here are two cool caterpillars. See how cleverly they use deception!
- The caterpillar of the Common Mormon Butterfly looks like bird poop.

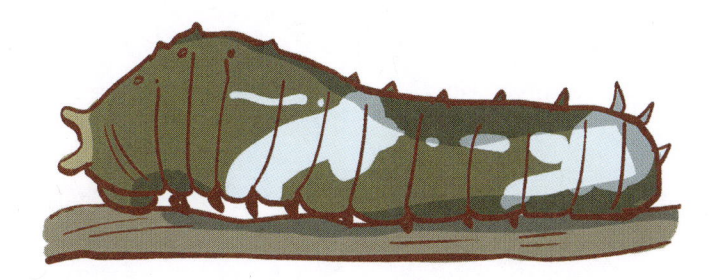

- The caterpillar of the Oleander Hawkmoth looks like vegetation and if a predator does happen to spot it, its large blue false eyes scare it away!

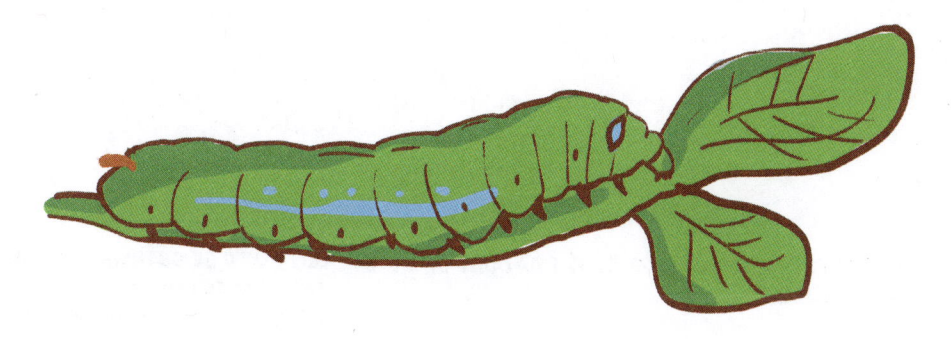

Recreate an alien using a caterpillar as inspiration.

ICYMI

The man who leaves no stone unturned is Karthikeyan S. He's a naturalist who lives in Bengaluru, and has written some amazing books about avenue trees, flowering shrubs and spiders. He's brimming with trivia about flora and fauna. Just be careful – if he tells you a fact once, expect to be quizzed about it the next time.

13 PEARLS OF THE QUEEN'S NECKLACE

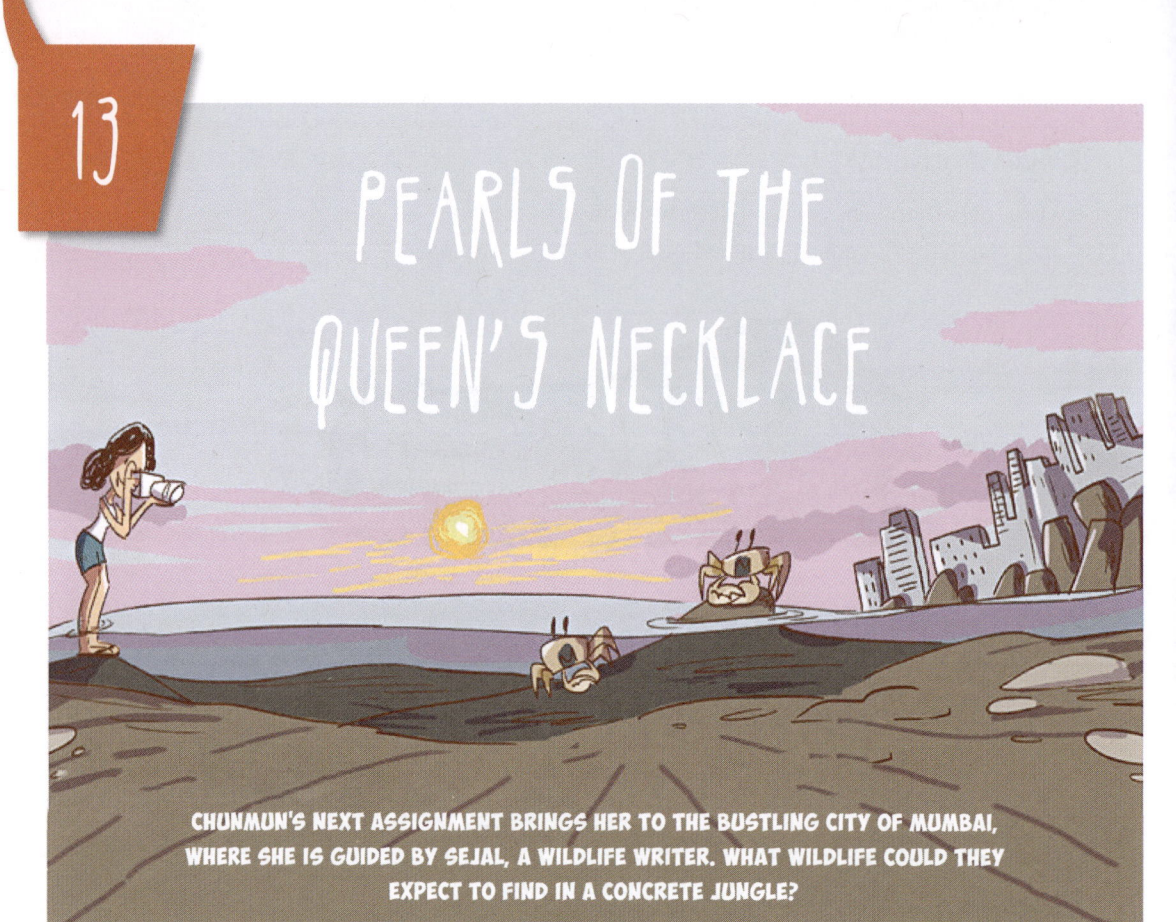

CHUNMUN'S NEXT ASSIGNMENT BRINGS HER TO THE BUSTLING CITY OF MUMBAI, WHERE SHE IS GUIDED BY SEJAL, A WILDLIFE WRITER. WHAT WILDLIFE COULD THEY EXPECT TO FIND IN A CONCRETE JUNGLE?

MY GOODNESS, SEJAL! LOOK AT THE AMOUNT OF LITTER ON THE BEACH!

THIS IS REALLY APPALLING, CHUNMUN.

BUT YOU'D BE SURPRISED TO KNOW THAT LITTER ISN'T THE ONLY THING MUMBAI'S BEACHES ARE STREWN WITH...

THIS ENTIRE STRETCH OF THE SHORE IN OUR LINE OF SIGHT IS ALSO TEEMING WITH WILDLIFE!

CAN YOU GUESS WHICH CREATURE HAS DOTTED THE BEACH WITH THESE TINY BALLS OF SAND?

UMM... A CRAB?

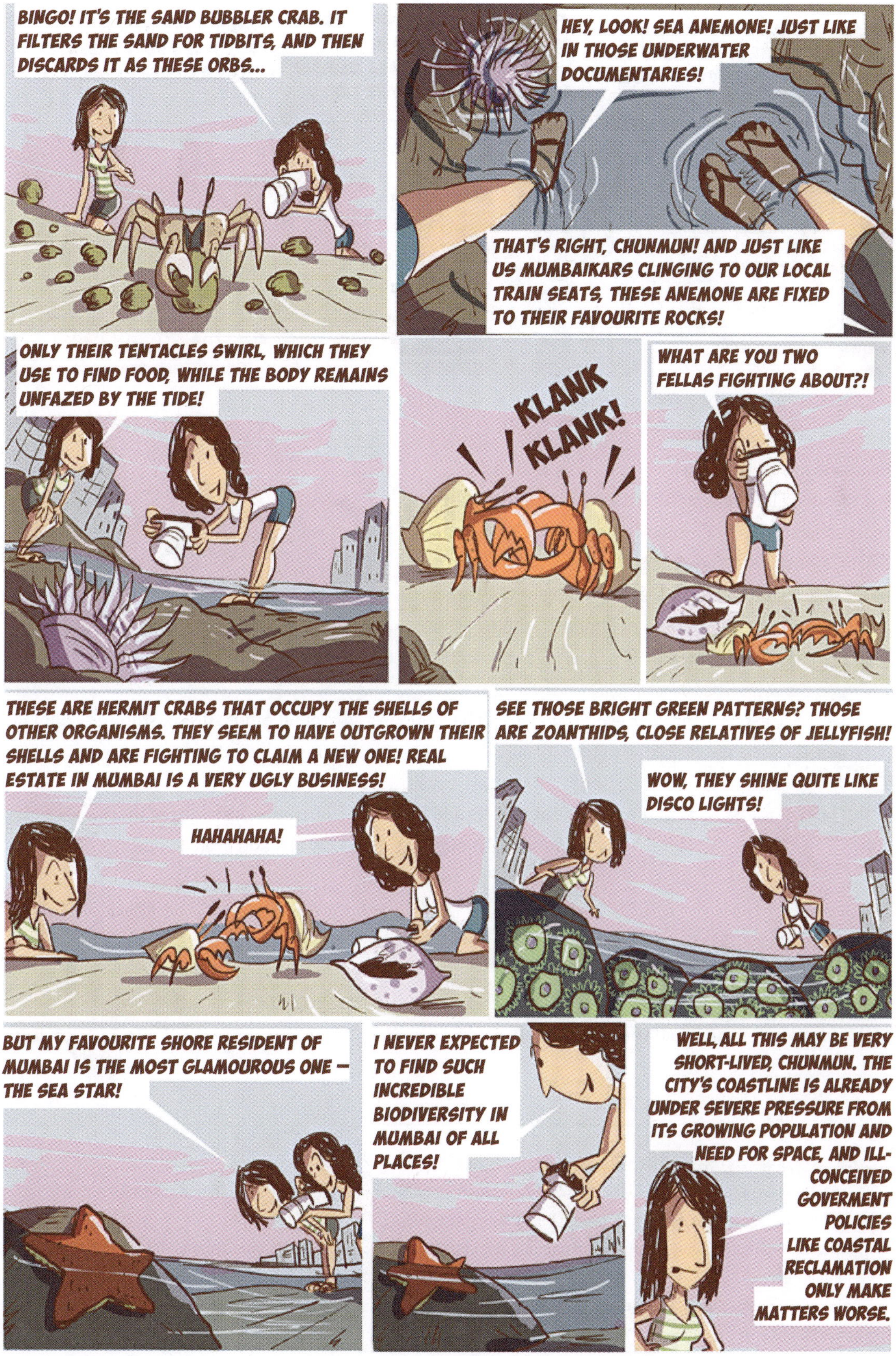
BINGO! IT'S THE SAND BUBBLER CRAB. IT FILTERS THE SAND FOR TIDBITS, AND THEN DISCARDS IT AS THESE ORBS...
HEY, LOOK! SEA ANEMONE! JUST LIKE IN THOSE UNDERWATER DOCUMENTARIES!
THAT'S RIGHT, CHUNMUN! AND JUST LIKE US MUMBAIKARS CLINGING TO OUR LOCAL TRAIN SEATS, THESE ANEMONE ARE FIXED TO THEIR FAVOURITE ROCKS!
ONLY THEIR TENTACLES SWIRL, WHICH THEY USE TO FIND FOOD, WHILE THE BODY REMAINS UNFAZED BY THE TIDE!
KLANK KLANK!
WHAT ARE YOU TWO FELLAS FIGHTING ABOUT?!
THESE ARE HERMIT CRABS THAT OCCUPY THE SHELLS OF OTHER ORGANISMS. THEY SEEM TO HAVE OUTGROWN THEIR SHELLS AND ARE FIGHTING TO CLAIM A NEW ONE! REAL ESTATE IN MUMBAI IS A VERY UGLY BUSINESS!
HAHAHAHA!
SEE THOSE BRIGHT GREEN PATTERNS? THOSE ARE ZOANTHIDS, CLOSE RELATIVES OF JELLYFISH!
WOW, THEY SHINE QUITE LIKE DISCO LIGHTS!
BUT MY FAVOURITE SHORE RESIDENT OF MUMBAI IS THE MOST GLAMOUROUS ONE — THE SEA STAR!
I NEVER EXPECTED TO FIND SUCH INCREDIBLE BIODIVERSITY IN MUMBAI OF ALL PLACES!
WELL, ALL THIS MAY BE VERY SHORT-LIVED, CHUNMUN. THE CITY'S COASTLINE IS ALREADY UNDER SEVERE PRESSURE FROM ITS GROWING POPULATION AND NEED FOR SPACE, AND ILL-CONCEIVED GOVERMENT POLICIES LIKE COASTAL RECLAMATION ONLY MAKE MATTERS WORSE.

> THIS INTERTIDAL ECOSYSTEM IS THE CITY'S BACKBONE. IT SUSTAINS OUR FISHING COMMUNITIES AND PROTECTS THE COASTLINE FROM FLOODING, EROSION AND NATURAL CALAMITIES.

> IT IS UPTO US COMMUNICATORS – PHOTOGRAPHERS AND WRITERS LIKE YOU AND I, TO MAKE MUMBAI REALIZE THAT THESE ARE THE REAL STARS THE CITY NEEDS...

> MUMBAI'S INTERTIDAL SUPERSTARS SURE DESERVE THEIR EXCLUSIVE PAPARAZZI!
>
> HAHAHA!

Here's the thing.

Everyone knows the Queen's Necklace in Mumbai. It's the stretch of Marine Drive in the southern part of India's financial capital which at night, when seen from a height, looks like the pearls in a necklace. Sadly, because of the coastal reclamation that's ongoing at the time of writing this book, the necklace has dimmed, as the coastline is being reworked and recast.

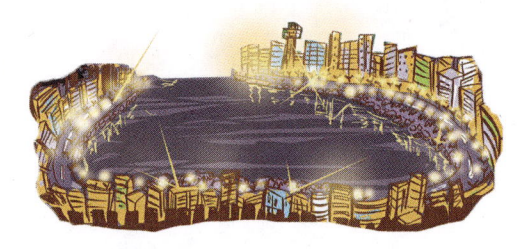

The real gems are the ones that are to be found in and around Mumbai's coastline as Chunmun discovered with the citizen-driven collective, Marine Life of Mumbai.

Yes, Mumbai has high-rises and Bollywood, but it also has some of the coolest tiniest fauna – the hermit crab, squids, shrimps, jelly fish and even octopi!

STARS AT YOUR FEET

If you're on a coastline, constellations of stars are of course in the sky, but also look closely on the sand, and you will find tiny sea stars as well.

FUN FACTS

1. Mumbai has a variety of intertidal habitats – rocky shores, sandy shores and mangroves.
2. Mumbai is home to over 50 species of crabs! Some of the more famous ones are ghost crabs, hermit and even fiddler crabs.
3. Zoanthids can feed both by photosynthesis and by capturing plankton!
4. Starfsh or Seastars have five arms on an average, but may even have six to eight arms! They are capable of regenerating arms lost in an injury!
Q. An intertidal creature that looks like a spiny ball is also found on the shores of Mumbai, and is a favourite prey of the starfish! Can you name this animal?

WHAT'S GOT EIGHT TENTACLES AND CAN GROW BACK AN ARM?

An octopus!

If you haven't watched it yet, check out the documentary film *My Octopus Teacher* by Pippa Ehrlich and James Reed. Set in South Africa's kelp forest, the filmmaker befriends a female octopus and what follows is truly magical.

WRITE A REVIEW OF THE FILM HERE:

Name of the film:
Director's name:
Year of release:
Actors/Voices:
Synopsis (Write in one paragraph what this film is about):

Tip: A good way to become an excellent reviewer is to read other reviews. You can easily find them online on news websites. Observe the writer's style. But don't copy!

What did you like about it?

What did you not like about it?

Conclusion:

NOT A FOREVER HOME

Hermit crabs cannot grow their own shells, which is why they go looking up and down the shore for discarded and abandoned shells. They need to be a perfect fit for them, just like how we all know our home is just the right one for us. And like a pair of jeans, they have to try out many shells before they find the right one. But these tiny adorable crabs often mistake plastic litter for a shell, like say a water bottle or a discarded doll's head, and may sometimes even die because they get trapped inside it! One study in the *Journal of Hazardous Materials*[1] estimated that in just one year, some 570,000 hermit crabs die in this manner.

LAUNDROMAT

At Juhu Beach, there's one tiny corner where you can plonk yourself down on the rocks and watch a whole colony of fiddler crabs. They're called that because their movement resembles the way a musician plays the fiddle, especially the movement of the male's one gigantic claw, which he's constantly waving. You can wave back, but it'll be futile. They're actually doing that to attract females.

PLASTIC EVERYWHERE!

Check out Bijal Vachharajani and Jayesh Sivan's book with WWF-India called *The Mystery of the Not-Missing Plastic*. It's free to read and download.[2]

It's a mystery where two teenagers investigate why is plastic everywhere in their lives. And it's definitely everywhere. Because, every day India produces almost 26,000 tonnes of plastic waste. That's 9.4 million tonnes a year. Only 5.6 million tonnes are recycled. A lot lands up on our shores when the sea spits it back at us.

BECOME A TIDE TURNER

You don't have to be Percy Jackson or sea god Poseidon to turn the tide on plastic. Join WWF-India and the United Nations Environment Programme's Plastic Tide Turner Challenge and become part of the solution. After all, people created the problem, we're going to have to be the ones to solve it. Check out the Tide Turners website[3] and join young people all across the world working to reduce plastic consumption.

EASY AS ABC

One simple thing to start doing to cut plastic from your life is to travel with a mini cutlery kit. Put together a small pouch of containing a fork, spoon and straw; pop it into your travel bag, and you will never need to use disposables again. Of course remember that forks and knives won't be allowed in flight hand luggage, so put that into your check-in baggage.

ICYMI

Sejal Mehta is part of the Marine Life of Mumbai collective where as a science communicator she inspires people to think about the city's shore. She's even written a book called *Superpowers on the Shore*. And best of all, she teamed up with Rohan Chakravarty to make picture books about the forest and its denizens.[4]

WHO DID THIS?

A B C

Can you identify who left this pattern behind?
1. Sand Bubbler Crabs
2. Snails
3. Wood Beetles

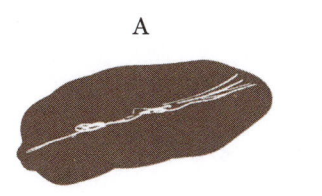

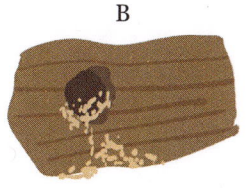

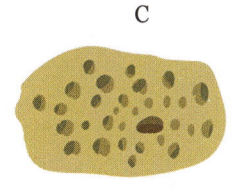

Answers: A-2, B-3, C-1

CRABBY ARTISTS

When walking on the beach if you suddenly come across a gorgeous radial pattern of sand balls, look carefully and you will spot a small hole in the middle. That's the home of the Sand Bubbler Crab that picks up sand in its moth, eats the plankton and detritus (yummy for them) and discards the rest, forming this gorgeous pattern.
Can you replicate it?

DRAW A FIDDLER CRAB

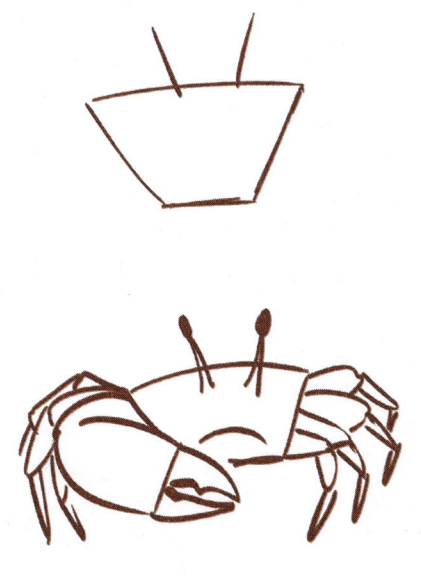

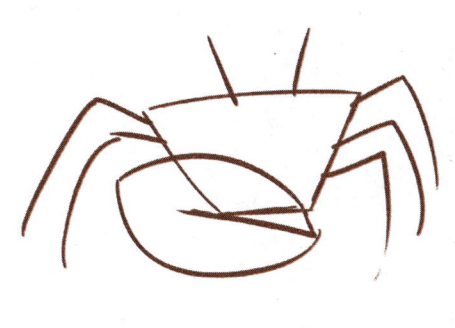

83

FALCON FLURRY

CHUNMUN TRAVELS TO A TINY VILLAGE IN NAGALAND TO SHOOT THE LARGEST CONGREGATION OF FALCONS IN THE WORLD! HERE SHE IS AIDED BY BANO, A JOURNALIST AND CONSERVATIONIST.

BANO, THIS IS THE FIRST VILLAGE I'M VISITING WHERE THERE ARE MORE BIRDS THAN PEOPLE... AND THAT TOO, BIRDS OF PREY!

YOU'RE RIGHT, CHUNMUN. NAGALAND IS A CRUCIAL PITSTOP FOR THE AMUR FALCON, THAT MAKES A RECORD-BREAKING MIGRATION FROM NORTH-EASTERN ASIA TO SOUTH AFRICA!

COME OCTOBER, AND MILLIONS OF THESE TINY BIRDS OF PREY VISIT THE VILLAGE OF DOYANG.

MILLIONS! THAT'S INCREDIBLE!

WHAT'S EVEN MORE ASTONISHING IS THAT THEIR ARRIVAL COINCIDES WITH THE EMERGENCE OF WINGED TERMITES — A VERY IMPORTANT SNACK THAT REFUELS THE BIRDS FOR THEIR ARDUOUS FLIGHT ACROSS THE ARABIAN SEA!

BUT NOT ALL WAS FUN AND FROLIC FOR THESE FALCONS. A FEW YEARS BACK, THESE HUNTERS BECAME THE HUNTED...

LOCAL FISHERMEN AND HUNTERS TRAPPED THOUSANDS OF THESE FALCONS IN THEIR NETS EVERY DAY, TO BE SOLD AS MEAT!

WITH THE HELP OF THE FOREST DEPARTMENT AND CONSERVATION GROUPS, WE HAVE SINCE BEEN ABLE TO ENFORCE A BAN ON HUNTING. BUT THAT ISN'T THE END OF THE PROBLEM, AS HUNTERS MUST HAVE AN ALTERNATIVE MEANS OF SUSTENANCE...

AND THIS IS WHERE PHOTOGRAPHERS LIKE YOU COME IN — TO POPULARIZE DOYANG AMONG BIRDWATCHERS AND TOURISTS THROUGH YOUR PICTURES!

YOU SEE, EVERY YEAR WHEN THE FALCONS ARRIVE, WE CELEBRATE THE AMUR FALCON FESTIVAL. MUSIC, DANCE PERFORMANCES AND EVEN FOOTBALL GAMES ARE CONDUCTED IN HONOUR OF THE WINGED VISITORS!

WELCOME BACK AMUR FALCONS

FRIENDS OF FALCONS

AMUR FALCON HOMESTAY

LONG LIVE AMUR FALCONS

FROM CHEAP BUSHMEAT TO BECOMING HONOURED GUESTS OF OUR VILLAGE, THE AMUR FALCON'S STORY HAS BEEN ONE OF THE GREATEST CONSERVATION SUCCESS STORIES IN RECENT TIMES!

WELCOME BACK AMUR FALCONS!

COME ALONG, CHUNMUN AND BANO! THE FESTIVITIES HAVE BEGUN!

A SIGHT, FANTASTIC

Imagine being in Doyang in Nagaland in the middle of October. It will be great weather most probably. After all it's early winter. But what's truly amazing is when you look up, the sky is filled with something. That's when you realized it's not something, but many birds. Many, many birds. It's the annual arrival of the Amur Falcons! No wonder, Nagaland is called the falcon capital of the world.

FUN FACTS

1. The Amur Falcon, despite being a bird of prey, is just 30 cms in length, about size of a woodpecker!
2. The male Amur Falcon is ashy-blue, while the female is grey with a creamy throat and belly, dotted with dark speckles.
3. Like most other raptors, female Amur Falcons are larger than males! This feature is called reverse sexual dimorphism.
4. The Amur Falcon's flight from the west coast of India to South Africa is the longest overwater migration for any bird of prey in the world.
Q. While termites feed Amur Falcons on their stopovers, a fast-flying, migratory insect is their chief prey item on their flight across the sea. Can you guess which kind of insect is this?

A WELCOME NOTE

When a Very Important Person visits your school, very often, someone gives them a welcome note. A speech to introduce them and greet them. Why don't you write one for the Amur Falcons as they touch down in India? You can quote other important people, explain why you think the birds are super cool, and maybe even add a rhyme or two?

....

....

....

HOP, SKIP AND FLY

The birds don't really stay in India, they only come for a quick pitstop – like when you stop in Lonavala for chikki and fudge while travelling from Mumbai to Pune. They snack on lots of juicy insects and fuel up for the rest of their flight. After all, flying takes energy, and what better than a protein-packed meal of grasshoppers?

THE NAME GAME

The word 'Amur' comes from a river in Russia and China. Apart from the Amur Falcon, there's the Amur Leopard and the Amur Tiger. There's also a pike, a catfish and a softshell turtle that have the word Amur in their name.

RIVER TALES

The word Amur is used in Europe, but
China calls it the 'Heilongjiang', or the Black Dragon River.

The Amur-Heilong River Basin is an amazingly well-preserved temperate forest.[1] But habitat loss, poaching and logging are threatening the animals there and their habitat.

But lots of organizations, including WWF, are working with local governments to help protect the landscape and its denizens.

FYI

You know how we take highways to reach different parts of India? Similarly, air routes or the flight paths taken by a migratory bird species is called a flyway!

DRAW

Draw an Amur Falcon in four easy steps.

REUSE, RECYCLE

The Amur Falcons are one clever species. They know much more about sustainability than all of us. Well, that's true of all the natural world. When it's time to build a nest, these falcons just find an old nest abandoned by another bird and move in. Now that's a good use of one of the 5Rs – Reuse.

ICYMI

Bano Haralu is a Nagaland-based journalist who is also a conservationist. Distressed for the Amur Falcon, she worked on making the hunting of the raptor illegal. She also worked closely with the community on boosting conservation efforts and on sustainable tourism.

MAP THE MIGRATION

Here's a map of Eurasia and Africa. Bano describes the Amur Falcon's migration from eastern Asia via India to Southern Africa and back, in the comic.

Can you draw this journey on the map?

15 HILLSIDE MUSE

With photography for her photo essay nearing completion, Chunmun visits the bank of the Himalayan river Tons in Garhwal, to gather some inspiration for writing her essay.

Uncle Bikky says that there's no better place than a hillside stream to sit and write...

But everything about this gorgeous river only distracts me!

This handsome Crested Kingfisher gulping down his eighth catch of the day...

That lively Brown Dipper taking dip after dip in the stream to hunt aquatic insects...

This dazzling Common Peacock butterfly mud-puddling...

Boy, I wish I weren't carrying my camera along! I'll never end up finishing my essay if all I do is shoot!

FOCUS, CHUNMUN, FOCUS!!
HUNDREDS OF EXPLORERS WRITE SUCH STUNNING PHOTO ESSAYS EVERY YEAR... I MEAN, LOOK AT ARATI'S WORK – HOW DOES SHE DO IT?! HOW DO I MAKE MY ESSAY STAND OUT?
THINK, CHUNMUN, THI--
-HUH!
CLICK
CLICK
CLICK!
SCRRRRRR!

Panel 1: (no text)

Panel 2: I'M SO SORRY THAT I SAID I SHOULD'VE LEFT YOU BEHIND, CAMERA!

Panel 3: NOW I KNOW EXACTLY WHAT WILL MAKE MY ESSAY STAND OUT.

NOT AN INSTA STORY

Before the world of Instagram posts and stories, there's always been the photo essay. If it sounds like something you have to do for homework, then worry not. Photo essays are not the stuff homework is made of.

Rather, they're an interesting visual deep dive into a subject.

You know how the saying goes, a picture is worth a thousand words? The photo essay as a medium combines an image with words to tell a complete story. Chunmun, being a wildlife photographer, often ends up shooting the images, and then writing an essay with them.

It sounds easier than it actually is. Photo essays can take years of patience and research. They of course can be of all types – they can be sad, funny, dramatic, black-and-white, colour, sepia.

So if you have access to a camera, (yes even a phone camera takes excellent photos now) and you are ready to fire-up your imagination, then the world's your oyster.

FUN FACTS

1. The Common Peacock Butterfly is Uttarakhand's state Butterfly!
2. The Dipper is the only songbird that is capable of swimming and diving!
3. Like other kingfishers, Crested Kingfishers can often be seen battering their prey on a rock before eating it. These birds eat upto their body weight's worth of fish every day!
4. The Tons river is the largest tributary of the Yamuna, which it meets near Dehradun, the capital of Uttarakhand.
Q. Another species of kingfisher that is coloured black-and-white quite like the Crested Kingfisher, is widespread across India, and is renowned for its hovering dives. Can you name this species?

BECOME A PHOTO ESSAYIST

Chunmun was waiting for inspiration to strike while creating her essay, and that's important. But you don't have to sit waiting around for inspiration. Here, it's not necessarily applicable that good things come to those who wait. You can get out and get inspired on your own.

Everything and anything can be the subject of a photo essay: When you think about it, the world's full of wondrous natural beings, and they're all fantastic photo essay subjects. Whether it's dew on a signature spider web or the back of a fern with its underside of spores, your family and friends (humans are animals too), or the tree outside your window, these are all possible subjects. All you need is a keen eye and a bit of imagination.

Be respectful and careful: Wildlife is fragile, and also potentially dangerous. Maintain a respectful distance from your subject. Don't disturb habitats. Don't take photos of predators, nests and babies that can put them or you in danger. Don't put yourself in harm's way. If you're taking a photo of a person, ask for permission first (we should be doing that for animals too!). You wouldn't like it if someone took a picture of you without checking with you, right? Nobody likes that annoying uncle at parties who forces everyone to stand and say famileeeee.

Research, research, research: Research your subject. If you are creating one about house geckos, then read up all about them, observe them, and that's when you will get better at understanding and appreciating them, and finally shooting and writing about them.

Try again: Don't just be happy with your first draft and first photo. Photographers take hundreds of photos before they are happy with one. In the same way, writers write a draft and keep refining it until they're happy with how it reads.

Choose wisely: Great, you have tonnes of photos now and you like all of them. Nope, not happening. Nobody wants to be like that friend who went to Maldives and took 562 photos of the shore and sun and now insists you see each and every one of them. You need to be ruthless and pick only the best ones.

Writing the essay: Like all stories, your essay also needs to have a beginning, middle and end. Introduce your essay, flesh it out and bring it to a conclusion. Most importantly, your picture is already doing a lot of the talking for you. So your essay needs to tell the reader something more – don't describe the photo saying a spider has eight legs or eyes. Instead, tell them little facts that you noticed while researching your essay, what were the thoughts flitting through your mind as the yellow butterfly danced in the sunlight as you captured its photo. Add facts, emotions, colour. Don't forget to proofread it.

Ask someone for advice: All photographers have photo editors and art directors, and writers have editors, just like you have a teacher who tells you how you did in your test. Once you have completed the photo essay, put it away for some time. Then take it out from the dark corner of your computer, desk and mind and show it to a trusted friend or adult who can give you good advice.

Let the world see your photo essay: You've worked so hard on your photo essay, so publish it online, print it out, contribute it as an entry to your school magazine, and share it with friends and family. Don't forget to add your name.

PHYSICS LESSONS FROM A KINGFISHER

Remember in Physics, when your teacher taught reflection and refraction? Or even if you don't, you have your text books to tell you what it is. So first, look it up.

Now look closely at the kingfisher that Chunmun saw. They are awesome fishing birds, but they owe their awesomeness to science.

Here's how: Have you gone splashing about in a pond or lake, trying to catch something in the water? Or even simply, tried to drop a coin in water and tried to catch it at first try? You won't be able to because the water refracts the object, making us perceive its position to be different. But the kingfisher has amazing ocular protection and they are fabulous at depth perception. Which means when they dive into water, they can perceive exactly where the fish is, and whoosh, they've caught their lunch.

Now that would be a cool thing to write about in a photo essay.

CREATE YOUR OWN PHOTO ESSAY

16 A BIRTHDAY SURPRISE

IT'S UNCLE BIKKY'S BIRTHDAY, AND CHUNMUN HAS JUST RETURNED FROM HER PHOTO EXPEDITION. SHE'S GOT A BIG SURPRISE FOR HER ORNITHOLOGIST UNCLE!

HAPPY BIRTHDAY, OLD PAL! I HAVE GIFTS FOR BOTH YOU BOYS!

OH BOY!

FIRST, SOME BRAND NEW CHEWBONES FOR DUGGU...

AND FOR YOU... TA-DAAA!

AH! THE LATEST ISSUE OF THE KINGDOM ANIMALIA EXPLORERS MAGAZINE!

NOT JUST THAT. THE LATEST ISSUE OF THE MAGAZINE WITH MY PHOTO ESSAY PUBLISHED IN IT!

OH MY MY! LET'S SEE...

THERE YOU ARE — PAGE 28...

Index
Editorial......1
The Magnolia, Molinillo and Colombia's favourite drink - Melisa Ayala..... 3
Romer's Tree Frog: A conservation story from Hong Kong - Alex Wong........ 18
The Great Indian Nature Trail - Chunmun Ganguly......28
Orca Dialects - Janie Wray.....30
Green Humour 38

96

MY GOODNESS, CHUNMUN, SUCH STUNNING PICTURES!
KINGDOM ANIMALIA
A GANGETIC DOLPHIN'S SNOUT AGAINST THE SUNSET!
A 'Blind' date
A LEAPING HOOLOCK GIBBON !
Mystical Mishmi
A LEOPARD'S PUGMARKS FACING A CAMERA TRAP! SPLENDID CONCEPT AND COMPOSITION!
Spots in the fields
AH, I LOVE THIS ONE! A THOUSAND AMUR FALCONS IN ONE FRAME! BRINGS BACK SO MANY MEMORIES OF NAGALAND!
Falcon Flurry!
HEY, WHICH BIRD IS THI--
KINGDOM ANIMALIA
Rising from the ashes!
Presumed extinct until this picture was
taken, the Himalayan Quail was last seen
145 years ago! This sighting of a mother
CHUNMUN!! YOU'VE DONE IN ONE YEAR WHAT I COULD NOT DO IN FORTY!
YOU'VE REDISCOVERED THE HIMALAYAN QUAIL!!
AND I'VE ALSO BEEN AWARDED A TRIP TO BORNEO FOR MY NEXT PHOTO ESSAY BY THE MAGAZINE!
MY GIRL, I COULD NOT BE ANY PROUDER!

THANK YOU, UNCLE BIKKY! I JUST HOPE THAT THE AUTHORITIES WILL READ MY ARTICLE AND ACT TO PRESERVE THE HABITAT OF THE HIMALAYAN QUAIL.

WHY, OF COURSE THEY WILL! I AM CALLING MY FRIEND, UTTARAKHAND'S CHIEF CONSERVATOR OF FORESTS RIGHT AWAY AND SHARING THE NEWS WITH HIM!

PSST, DUGGU, DID YOU NOTICE HOW UNCLE BIKKY HAS COMPLETELY FORGOTTEN ABOUT HIS ARTHRITIS? HE'S SPRINTING LIKE THE QUAILS IN MY PICTURE!

DISCLAIMER: As of 2024, the Himalayan Quail is listed as 'Critically Endangered' on IUCN's Red List and is presumed to be extinct by ornithologists. The rediscovery of the bird has been staged in the story for creative purposes, and is purely fictional.

Before you go leaping off like Uncle Bikky, you should know that the rediscovery of the Himalayan Quail hasn't happened, and this is a fictionalized conversation. Wouldn't it be amazing if an extinct bird was to come back? But sadly, because of many reasons – poaching, habitat loss, climate change – extinction is getting more and more common.

Global warming could render some one-third of Earth's plants and animals extinct by 2050.[1] That is not as far away as it sounds. We're making the planet inhospitable for all these beings. And let's not forget how much climate change impacts us.

BUT

Chunmun has something very important at her disposal. It's the power of the written word and the visual image – the media.

Here are some cool careers in media where people are working to spread the word about wildlife and climate change.

FILMMAKERS

Film is perhaps one of the most powerful ways to communicate how awesome the natural world is, as well as some of the threats it's facing. Images, writing and music and commentary come together to create some really impactful movies and documentaries.

- Most wildlife people will readily tell you that British broadcaster and biologist David Attenborough has been a huge inspiration for them. Check out his *Planet Earth* and *Blue Planet* films.
- Mike Pandey's film *Shores of Silence* led to the ban on the fishing of whale sharks and to listing them as endangered.
- Rita Banerji not only makes films, but she also started Green Hub India to mentor aspiring filmmakers in the North-east and central India.

EDITORS

When you read any well-written story, you should know there's an invisible person behind it. That's the editor, who often commissions the story, works on shaping it for the reader and publishes it.

- Bittu Sahgal started India's first wildlife magazine, *Sanctuary Asia* and its sister publication, *Sanctuary Cub* for young readers. Bittu also started Kids for Tigers, a national programme that gets children to work together to protect the tiger.
- Kartik Chandramouli is the digital content editor of *Mongabay India*, which has some of the best reportage on wildlife and science.
- A filmmaker and photographer, Radha Rangarajan is the editor at *Nature InFocus*, a website that focuses on bringing together communities of wildlife storytellers.
- Megha Moorthy is the editor-in-chief of *RoundGlass Sustain*, a website that tells stories about Indian wildlife.
- An evolutionary ecologist and author, Kartik Shanker does many things. He is the founder editor of *Current Conservation* magazine.

REPORTERS AND JOURNALISTS

Wildlife and environment reporters tell the world the stories of what's happening to India's natural world.
- Prerna Singh Bindra has been a journalist, written several books on wildlife and is a conservationist who has a soft spot for elephants.
- Swati Thiyagarajan has been a broadcast news journalist and is now a documentary filmmaker.
- Bahar Dutt has been a TV news journalist, and is also a teacher and conservationist.
- Janaki Lenin writes about science and wildlife, and has written many books about them.
- Neha Sinha has been a reporter, a conservationist and has trained many young people on policy.
- Jay Mazoomdar is perhaps one of the foremost voices in wildlife reporting, including a story that brought to light India's tiger crisis.

AUTHORS

Without this group of writers, you'd have no books to read. Children's book authors imagine stories, crunch non-fiction trivia and tell them in a fun way, bringing together humour, drama, emotions to make young readers wonder about the natural world.
- Ranjit Lal writes stories with his tongue firmly in his cheek. His books are funny and adorable. Check out *The Tigers of Taboo Valley* about a tiger daddy who finds himself taking care of his cubs (a rarity in real life).
- Meghaa Gupta's *Unearthed* tells the environmental history of India.
- Zai Whitaker's stories come from a place of knowing, having worked in wildlife for decades, and she also co-founded the Madras Crocodile Bank Trust. Her book *Salim Mamoo and Me*, illustrated by Prabha Mallya is an adorable insight into the life of a girl who grows up in a family that loves birds.

ACTORS AND PERFORMERS

With millions of people watching films and following actors on social media, when some of these stars champion wildlife and raise their voices to highlight critical issues, they can inspire others to join conservation efforts.
- Bollywood actor Dia Mirza's work revolves around amplifying her voice for India's wildlife and all things green.
- Hollywood actor Leonardo DiCaprio's character may not have got saved in the Titanic, but he's trying to turn the tide around climate change and has even made a film *Before the Flood*.
- Hollywood actor Mark Ruffalo has a non-profit called *The Solutions Project* which works against climate change.
- Music composer Ricky Kej creates works that centre around the environment. He's even won Grammy awards for his music!

PHOTOGRAPHERS

Photography is perhaps one of the biggest ways in which people learned about the splendours of the natural world without having to travel anywhere.
- We've discussed this already. There are some useful tips in Chapter 15 as well.

ARTISTS

From the time human beings started telling stories, our ancestors have been drawing images on cave walls. Artists, illustrators and cartoonists wield their pencils like magicians' wands to create magic on paper and screen.
- Odiya artist Sudarshan Shaw brings together indigenous art, communities and science in his art work.
- Rohan Dahotre's bold illustrations celebrate nature and wildlife.
- Sangeetha Kadur's beautiful paintings will make you feel like you are in that habitat!
- Nirupa Rao is a botanical illustrator whose work is a deep dive into flora and habitats.

STAY INFORMED

Read, watch and listen when it comes to wildlife, when you can't go experience it first-hand. These are just a handful of people doing some amazing work. We're sure you know lots more.
Make a list of your favourite media here.

My favourite books:

My favourite films:

My favourite songs:

....

....

....

....

My favourite magazines:

....

....

....

....

A photograph I loved:

....

A story I liked reading:

....

WELCOME HOME!

What a whirlwind journey it's been with Chunmun. Isn't India just awesome, with all her magnificent habitats from forests to mangroves to wetlands to deserts, and the species that call them their home?

By travelling virtually, you have saved many carbon miles, so here, take a moment to pat yourself on the back for that as well.

But our country's precious wildlife is facing many threats. To begin with, climate change is revving up extinction at an alarming rate. Throw in deforestation, poaching, habitat loss, pollution and it's a recipe for disaster.

If you're wondering what you can do sitting in your house, then there's plenty!

Start by understanding the natural ecology and biodiversity of your habitat – that's your home. Keep a nature journal. WWF-India's Biodiversity Explorer Kit will help you get started.[2]

Small individual and collective actions go a long way for the planet. Act local, think global – that makes you a glocal citizen.

Remember the 5 Rs?

Reduce

Reuse

Recycle

Review

Refuse

Get cracking on them.

Get more people involved. The more the merrier.

Be a responsible tourist. You already know how.

Be a responsible citizen. Grow up and vote for parties that are committed to green policies. After all, if we don't have policies that are about fresh air, clean drinking water and rich soil, how are we to live?

Put your creative skills to good use. Whether it's designing a button for the Fishing Cat, drawing a poster for the Bugun Liocichla, developing a photo essay about fragmented habitats, performing a rap song or a play about snakes, taking a picture of a spider, drawing a cartoon about waste management, it's all important. Don't forget to share it with family and friends.

Observe WWF's Earth Hour. You get to be part of a global movement on the last Saturday of the month of March. Switch off all non-essential electric lights for one hour from 8–9pm. In 2024, people from over 180 countries and territories participated in Earth Hour. They didn't just switch off lights, they did mangrove planting drives, had virtual concerts (in the pandemic), cleaned-up their streets and a lot more. That one hour is meant to make people think about energy usage and renew their commitment to their home, planet Earth. And of course, it's not limited to one hour in the year. DUH! It's about our commitment to the planet and its denizens every single day. Because it's our only home. There is no planet B.

ANSWERS TO FUN FACTS

Comic 1 – Underground stem/rhizome

Comic 2 – Arunachal Macaque

Comic 3 – Takin

Comic 4 – Irrawaddy Dolphin

Comic 5 – Musk Deer

Comic 6 – Amur Leopard

Comic 7 – White-throated Kingfisher

Comic 8 – Hume's Warbler

Comic 9 – Mugger

Comic 10 – Common Bottlenose Dolphin

Comic 11 – Caecilians

Comic 12 – The Common Mormon Butterfly mimics the Crimson Rose Butterfly, the Common Palmfly mimics the Plain Tiger Butterfly, the Handmaiden Moth mimics wasps, Robberflies mimics bees and so on.

Comic 13 – Sea Urchin

Comic 14 – Dragonfly (more specifically, the Globe Skimmer or Wandering Glider)

Comic 15 – Pied Kingfisher

NOTES

CHAPTER 1: FROM THE NORTH-EAST, WITH LOVE

1. University of Western Australia, 'Study Reveals Plants "Listen" to Find Sources of Water', *phys.org*, 11 April 2017, https://phys.org/news/2017-04-reveals-sources.html#jCp.

CHAPTER 2: BUGUN IN THE BUSH

1. University of East Anglia, 'Indigenous and Local Communities Key to Successful Nature Conservation', https://www.sciencedaily.com/, 2 September 2021, https://www.sciencedaily.com/releases/2021/09/210901191428.htm.

2. Niranjan Kaggere, 'Karnataka: New Freshwater Crab Species Spotted in Western Ghats', *Times of India*, 20 August 2022, http://timesofindia.indiatimes.com/articleshow/93669550.cms?utm_source=contentofinterest&utm_medium=text&utm_campaign=cppst.

3. T.V. Jayan, Indian Herpetologists Discover New Viper Species in Arunachal Pradesh, *The Hindu*, 23 April 2020, https://www.thehindubusinessline.com/news/variety/indian-herpetologists-discover-new-viper-species-in-arunachal-pradesh/article31403574.ece.

4. Sandhya Ramesh, 'This Researcher Has Named His Latest Find after Modi Govt's Top Science Official', *The Print*, 26 December 2018, https://theprint.in/science/this-researcher-has-named-his-latest-find-after-modi-govts-top-science-official/168953/.

5. https://xeno-canto.org/

6. Shreya Dasgupta, 'From a New Bird to a Community Reserve, Arunachal's Bugun Tribe Sets Example', *The Wire*, 29 December 2018, https://thewire.in/environment/arunachal-bagun-liocuchla-eaglenest-wildlife-sanctuary.

CHAPTER 3: MYSTICAL MISHMI

1. Shri Jim Pulu, *A Handbook on Idu Mishmi Language*, Government of Arunachal Pradesh, February 2002, http://www.rogerblench.info/Language/NEI/Mishmi/Idu/Idulang/Idu%20Mishmi%20handbook%20Pulu%202002.pdf.

2. Nature in Focus, 'All About the Himalayan Monal', https://www.natureinfocus.in/, 12 October 2021, https://www.natureinfocus.in/animals/all-about-the-himalayan-monal.

3. Sangeeta Yadav, 'These Wild Men', *The Pioneer*, 5 November, 2017, https://www.dailypioneer.com/2017/sunday-edition/these-wild-men.html.

CHAPTER 4: DOLPHIN DATE

1. Bijal Vachharajani, 'Sir, over 60 scientists have . . . dispelling myths about them.' [X Post], 11:32 a.m., https://twitter.com/, https://twitter.com/bijal_v/status/1254289730562211842?s=20.

2. Andrea Silen, '5 Reasons Why Bats Are the Best', https://kids.nationalgeographic.com/, https://kids.nationalgeographic.com/explore/5-reasons-why-hub/5-reasons-bats-are-best/.

3. WWF, 'Ganges River Dolphin', https://www.worldwildlife.org/, https://www.worldwildlife.org/species/ganges-river-dolphin.

4. Wildlife Institute of India, 'Indian Tent Turtle', https://wii.gov.in/, 27 July 2023, https://www.wii.gov.in/nmcg/priority-species/reptiles/indian-tent-turtle.

5. Nature, 'What I Learnt Pulling a Straw Out of a Turtle's Nose', https://nature.com/, 6 November 2018, https://www.nature.com/articles/d41586-018-07287-z.

6. Aradhana Wal (Curator), Bijaya Das (Editor), 'India Produces over 25,000 Tonnes of Plastic Waste a Day: Environment Ministry', https://www.news18.com/, 30 December, 2017, https://www.news18.com/news/india/india-produces-over-25000-tonnes-of-plastic-waste-a-day-environment-ministry-1618383.html.

7. Arati Kumar-Rao, 'About & Contact', https://www.aratikumarrao.com/, https://www.aratikumarrao.com/about.

CHAPTER 5: DEER, DACHIGAM

1. Harini Nagendra, 'What Hyder Ali and Tipu Sultan Had to Do with Bangalore's Love Affair with Trees', https://scroll.in/, 2 February 2017, https://scroll.in/article/828228/what-hyder-ali-and-tipu-sultan-had-to-do-with-bangalores-love-affair-with-trees.

2. Sanchari Pal, 'How Bengaluru Went from Being a Barren Plateau to a Verdant Garden City in 250 Years', https://www.thebetterindia.com/, 17 February 2017, https://www.thebetterindia.com/87537/barren-green-history-bengaluru-vijay-thiruvady/.

3. Prerna Singh Bhindra, 'Nazir Malik and Tahir Shawl', https://sanctuarynaturefoundation.org/, https://sanctuarynaturefoundation.org/award/nazir-malik-and-tahir-shawl.

CHAPTER 6: SPOTS IN THE FIELDS

1. National Geographic, 'First Mammal Species Goes Extinct Due to Climate Change', https://www.nationalgeographic.com/news/2016/06/first-mammal-extinct-climate-change-bramble-cay-melomys/.

CHAPTER 7: THE FISHER FROM BENGAL

1. *catinwater*, 'Fishing Cat Mystrey Question: And the Answer is . . . (Photos)', https://catinwater.wordpress.com/, 11 December 2011, https://catinwater.wordpress.com/2011/12/11/fishing-cat-mystery-question-and-the-answer-is-photos/.

2. Verity Mathis, 'What Can You Learn from Studying an Animal's Scat?', https://theconversation.com/global, 25 November 2019, https://theconversation.com/what-can-you-learn-from-studying-an-animals-scat-126307#:~:text=Scat%2520can%2520tell%2520us%2520a,clues%2520about%2520the%2520animal's%2520diet.

3. National Centre for Biological Sciences, 'What Makes the Black Tigers of Simlipal Black?', https://www.ncbs.res.in/, 14 September 2021, https://www.ncbs.res.in/content/what-makes-black-tigers-similipal-black.

CHAPTER 8: WARBLER AT THE WINDOW

1. Ed Stubbs, 'Focus on Greenish Warbler', https://www.birdguides.com/, 24 August 2021, https://www.birdguides.com/articles/species-profiles/focus-on-greenish-warbler/.

2. Bird Count India, 'Green & Greenish Warbler', https://birdcount.in/, https://birdcount.in/migration-map/grewar3-grnwar1/.

3. Byju's, 'List of Indian State Birds', https://byjus.com/, https://byjus.com/free-ias-prep/state-birds-of-india/.

4. Clay Bolt, '3 Things You Can Do to Help Your Local Pollinators', https://www.worldwildlife.org/, 24 June 2021, https://www.worldwildlife.org/stories/3-things-you-can-do-to-help-your-local-pollinators#:~:text=Everyone%2520knows%2520the%2520honey%2520bee,the%2520US%2520and%2520Canada%2520alone%253F.

CHAPTER 9: RAVINE REPTILIAN

1. Dharmendra Khandal, 'Discovery of New Gharial Population', https://www.saevus.in/, 6 March 2018, https://www.saevus.in/chambal-river-the-last-bastion-for-the-gharial-population/.

2. Ramki Sreenivasan, 'Gharials on the Chambal', https://www.conservationindia.org/, 3 October 2014, https://www.conservationindia.org/gallery/gharials-on-the-chambal.

3. Wildlife Institute of India, 'Red Crowned Roofed Turtle', https://wii.gov.in/, 27 July 2023, https://www.wii.gov.in/nmcg/priority-species/reptiles/red-crowned-roofed-turtle.

CHAPTER 10: THE HUMPBACK OF THE ARABIAN SEA

1. Bhavya Dore, 'Did You Call? Misuse of Bird Call Audio is Disturbing Bird Behaviour', https://india.mongabay.com/, 26 september 2019, https://india.mongabay.com/2019/09/did-you-call-misuse-of-bird-call-audio-is-disturbing-bird-behaviour/#:~:text=Going%2520by%2520the%2520law%252C%2520using,on%2520the%2520species%2520being%2520studied.

2. Open Axis, 'Bringing Science and Activism to Your Goan Holiday: Puja Mitra, Ethical Marine Tour Guide', https://openaxis.in/, 7 November 2021, https://openaxis.in/2021/11/07/bringing-science-and-activism-to-your-goan-holiday-puja-mitra-ethical-marine-tour-guide/.

CHAPTER 11: WHERE FROGS FLY!

1. Dr Seshadri K. S., 'High Life: Gliding Frogs of the Western Ghats', https://roundglasssustain.com/, 24 September 2023, https://roundglasssustain.com/species/high-life-gliding-frogs-western-ghats.

2. Ht Weekend, 'Photos: Out in the Field with the Frog Prince of India', *Hindustan Times, 16 July 2021*, https://www.hindustantimes.com/ht-weekend/photos-out-in-the-field-with-the-frog-prince-of-india-101626363177688.html.

CHAPTER 12: MACRO MAGIC!

1. Karthikeyan S., 'Owlfly', https://jlrexplore.com/, 1 December 2020, https://jlrexplore.com/explore/naturalist-s-corner/owlfly.

2. Geetha Iyer, 'Netwinged Insects: The Invisible Caretakers of Forests', https://india.mongabay.com/, 19 July 2018, https://india.mongabay.com/2018/07/netwinged-insects-the-invisible-caretakers-of-forests/.

CHAPTER 13: PEARLS OF THE QUEEN'S NECKLACE

1. Sophie Lewis, 'Plastic Pollution has Killed half a Million Hermit Crabs That Confused Trash for Shells', https://www.cbsnews.com/, 7 December 2019, https://www.cbsnews.com/news/plastic-pollution-has-killed-half-a-million-hermit-crabs-that-confused-trash-for-shells/.

2. Bijal Vachharajani, Jayesh Sivan, 'The Mystrey of the Not-Missing Plastic', https://tide-turners.org/, 2021, https://tide-turners.org/assets/Files/The%20Mystery%20of%20the%20Not-Missing%20Plastic%20ENGLISH%20SR.pdf

3. https://tide-turners.org/.

4. http://www.storyweaver.org.in.

CHAPTER 14: FALCON FLURRY

1. WWF, 'Amur-Heilong River Basin Reader', https://wwf.panda.org/, February 2008, https://wwf.panda.org/discover/knowledge_hub/where_we_work/amur_heilong/.

CHAPTER 16: A BIRTHDAY SURPRISE

1. Center for Biological Diversity, 'Global Warming and Endangered Species Initiative', https://www.biologicaldiversity.org/, https://www.biologicaldiversity.org/campaigns/global_warming_and_endangered_species/index.

2. https://academy.wwfindia.org/biodiversity-explorer-kit/.

A NOTE ON THE AUTHORS

ROHAN CHAKRAVARTY is a cartoonist, illustrator and the creator of *Green Humour*, a series of cartoons, comics and illustrations on wildlife and nature conservation. Cartoons from *Green Humour* appear periodically in newspaper columns, magazines and journals. Illustrations from *Green Humour* have been used for several projects and campaigns on wildlife awareness and climate change. Rohan is also the author of seven books (including *Green Humour for a Greying Planet*, *Naturalist Ruddy* and *Pugmarks and Carbon Footprints*) and has won awards by UNDP, Sanctuary Asia, WWF International, the Royal Bank of Scotland and Bangalore Literature Festival for his work. He is notorious for rolling up into a ball like a pangolin to avoid answering the phone or meeting people.

When BIJAL VACHHARAJANI is not reading a children's book, she's writing or editing one. She's the author of multiple planet-friendly books including *A Cloud Called Bhura*, *Savi and the Memory Keeper* – both of which won the AutHer Award, and *When Fairyland Lost its Magic*, which won the Kalinga Award. A commissioning editor at Pratham Books, she has been a journalist, an animal rescuer and managed a magazine. Now she's mostly a climate worrier.

ABOUT INDIAN PITTA KIDS

Indian Pitta Kids is India's first dedicated children's imprint about nature and wildlife, published in association with WWF-India. Our books will take young readers on exhilarating, informative and funny journeys into the world of the wild.

ABOUT INDIAN PITTA

Indian Pitta Kids is part of the Indian Pitta imprint. Our books about birds and natural history go beyond field/identification guides, to explore the bigger mosaic of habitats, ecosystems and human interactions that touch the lives of birds. Successful conservation programmes, troubling environmental challenges, personal exploration of a landscape, deep dives into the ecology of a species, the quest for a rare species and the sheer joy of birding – these are some of the ideas that you can expect to explore within the pages of our books.

ABOUT WWF-INDIA

WWF-India is registered as a Public Charitable Trust. We are an environmental organization with extensive on-ground experience. We combine this knowledge with scientific research and practical insights to create solutions. Our conservation approach is holistic and integrated, connecting wildlife, communities, natural habitats, governments and corporations. This interconnectedness is what makes us unique.